Farm health and productivity management of dairy young stock

Farm health
and
productivity management
of dairy young stock

Siert-Jan Boersema

Joao Cannas da Silva

John Mee

Jos Noordhuizen

Wageningen Academic
P u b l i s h e r s

ISBN: 978-90-8686-129-3
e-ISBN: 978-90-8686-694-6
DOI: 10.3920/978-90-8686-694-6

Photo cover by courtesy of John Mee

First published, 2010

© Wageningen Academic Publishers
The Netherlands, 2010

This publication and any liabilities arising from them remain the responsibility of the authors.

Contents

List of acronyms and short explanation of some technical terms

BCS	body condition score (an estimate to score muscle and fat coverage of the animal's body at the lumbar and tail head area)
BRD	bovine respiratory disease
Business unit	part of the whole chain of on-farm production process elements (for example: the colostrum period; or: the pasture period)
BVD	bovine virus diarrhoea
BW	body weight
C/B ratio	costs-to-benefits ratio
CCP	critical control point (in quality risk management, QRM)
DM	dry matter
FAO	food and agriculture organisation of the United Nations, Rome, Italy
FHPM	farm health and productivity management (programmes); highly comparable to HHPM, but addressing the farm as a whole, including the animals, their environment, and the management
GDF	good dairy farming (code of practice)
HACCP	hazard analysis and critical control points
HHPM	herd health and productivity management (programmes); mainly focusing on the 'herd' as unit of concern
IBR	infectious bovine rhinotracheitis
IgG	immunoglobulin type G
IM	intramuscular (injection)
IRR	internal rate of return
IU	international units
IV	intravenous (injection)
NPV	net present value
OIE	Office Internationale d'Épizooties, Paris, France
OR	odds ratio (an epidemiological risk quantification parameter: OR >1 means risk increase, OR < 1 means risk decrease, OR= 1 means no association between factor and disorder)
PI	persistently infected animal (BVD)
POPA	point of particular attention (in quality risk management, QRM)
PPM	parts pro million
QRM	quality risk management (programmes), addressing animal health, animal welfare and public health issues at farm level
SC	subcutaneous (injection)
SMART	specific-measurable-achievable-relevant-time based (method to raise focussed questions and to set operational goals, e.g. in FHPM)
TQM	total quality management

Work instruction   practical, technical instruction sheet for clarifying activities in a certain, specific area (for example, hygiene and disinfection of calf houses)

Preface

The rearing of young stock until calving on dairy farms is often neglected, as compared to the management of adult – lactating – cattle.

Some farmers believe that calves and maiden heifers are non-productive, for at least two years. Others say they can not do without them because they will sooner or later replace older cows that will be culled some day.

It is often not realised that young stock represent a critical investment in the future of the dairy farm and this should be a reason to pay sufficient attention to the rearing period to safeguard that investment and get the best efficiency out of it. *Productivity* in the rearing period can be regarded as the optimal weight gain during the whole rearing period, paralleled by an optimal animal health state, given the herd health state, and an optimal reproductive potential. The ultimate goals of this rearing are the heifer after its first calving entering into milk production up to her genetic potential, and that she will show an optimal health and longevity in the herd.

This book has been designed to provide the reader (practising veterinarians, herd health veterinarians, extension officers and other farm advisors, but also dairy farmers) with *practical tools* to manage the whole complexity of young stock rearing.

We have chosen a *population-medicine approach* because it is with a herd of healthy animals that the farmer earns his income, and not with individual, diseased ones which cost him money. This approach provides the farmer with structure, planning, evaluation, organisation, and coaching.

Risk identification and risk management are key aspects of the population medicine approach, and therefore the Annexes comprise a series of practical and useful worksheets.

Ultimately, a section on diseases in young stock which affect the herd rather than just individuals is added as a quick reference guide because they have become rather complex and are economically relevant to the farmer. Nevertheless, this book should not be regarded as the textbook on calf diseases; other, more comprehensive books are available.

We hope that this text will be a beneficial tool in your daily practice, and, if you are currently not approaching young stock problems using a population-medicine methodology, that this book will assist and encourage you to do so. Young stock rearing is about investing in the future of the farm by both the farmer and the veterinarian. Optimising herd health and production management of dairy young stock means optimising performance, and at the same time, reducing losses and costs, and, hence, increasing animal welfare and farm profitability. Veterinarians have a substantial role to play.

January 2010
The authors:

Siert-Jan Boersema (the Netherlands)
    Veterinary Practice '*Van Stad tot Wad Dierenartsen*', Loppersum, the Netherlands
    Institute for Risk Assessment Studies IRAS, Faculty of Veterinary Medicine, Utrecht University, the Netherlands
    VACQA-International, Santarém, Portugal

Joao Cannas da Silva (Portugal)
    Facultade de Medicina Veterinaria, Universidade Lusofona, Lisboa, Portugal
    VACQA-International, Santarém, Portugal

John Mee (Ireland)
    Teagasc, Moorepark Dairy Production Research Centre, Fermoy, Ireland

Jos Noordhuizen (France)
    VACQA-International, Santarém, Portugal
    Associé à l'École Nationale Vétérinaire de Lyon, France

Photos in this book are reproduced by courtesy of John Mee, Joao Cannas da Silva, Siert-Jan Boersema and Jos Noordhuizen

Section I. The population medicine approach to young stock rearing

Chapter 1. Economic aspects of young stock rearing

(estimation of disease losses, and the cost-benefit assessment of intervention measures)

1.1. General issues

The rearing of young stock (replacement heifers) on a dairy farm is a costly business without instant revenues. Not only because it takes, by-and-large, two years to raise a heifer until her first calving, but also because during that period different diseases and disorders may occur which all have an economic impact and cause (more or less considerable) economic losses (Dirksen *et al.*, 1984; Noordhuizen, 2004).

Young stock rearing is – from an economic and enterprise point of view – a specific component of dairy farming, a particular farm process, next to cow husbandry and feed harvesting, pasturing and feeding management (Figure 1.1). Young stock rearing can be considered as an investment in the future, while at the same time the revenues are rather indirect. These revenues as features of productivity are, for example, an optimal growth pattern throughout the rearing period, animals in good health and a good genetic make-up.

In order to be able to make estimations on economic losses, or on costs and benefits, we need information, such as about disease prevalence, the variation between farms, and price elements. And even if we have that information, the outcome of our estimations is not always the same. The latter is caused by the fact that the effects of disease are not always clear and manifest, that these effects show a temporal distribution, and that

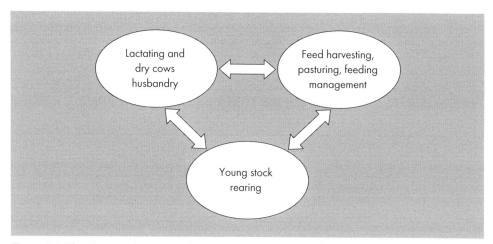

Figure 1.1. The three major economic components of a dairy farm.

these effects are affected by other factors such as nutrition, barn climate or housing conditions. This multifactorial nature of disorders is a major reason to implement a population-medicine approach to young stock rearing (Noordhuizen *et al.*, 2004).

1.2. Disease losses

There are different elements to consider when assessing economic loss due to diseases or disorders. The major elements are listed in Table 1.1. Of the elements listed in Table 1.1, the three most relevant ones are: productivity loss; reduced slaughter or market value; and premature culling (Huirne *et al.*, 2002).

International literature is not very abundant regarding the estimation of economic losses due to the different diseases during the period of young stock rearing. This means that we can only address some rearing issues in a fragmentary way.

1.2.1. The costs of rearing young stock

Rearing costs can be considered as the second largest production cost component on a dairy farm (Cady and Smith, 1996). This should be reason enough to pay sufficient attention to young stock rearing, and its associated costs.

Table 1.1. Overview of major elements concerning the economic losses due to diseases in dairy cattle.

Major elements comprised in economic losses due to disease in dairy cattle
Costs related to:
treatment (medicinal products)
discarded milk due to treatment (with e.g. antibiotics)
extra labour by farm worker(s) and reduced productivity and labour satisfaction
retarded growth in calves and maiden heifers
delayed age at first calving in heifers
loss of body condition and body weight (loss in carcass value)
longer dry periods
higher mortality
premature culling and lower cull animal value
loss of milk production or deviating quality
loss of genetic potential
loss of production potential (due to lung or udder tissue problems)
diagnosis (veterinarian; laboratory testing; *post-mortem*)

The different countries have different price systems and are, hence, not really comparable. Based on generally accepted target values for rearing (e.g. pre-weaning death rate of calves <10%; first age at calving approximately 25 months; herd culling rate 25%; average herd calving interval <13 months), we can make an estimation of those costs, or changes in those costs when the herd performance deviates from the given target value.

A study by Tozer and Heinrichs (2001) showed that rearing costs may be around $32,000 (≈ €26,600) for a 100 cow herd. If herd culling rate increased from 25% to 30%, there would be too few heifers to replace cows. If, however, herd culling rate dropped from 25% to 20%, the rearing costs would decrease by about $8,000 (≈ €6,600 or 24.6%). And if the age at first calving dropped from 25 to 24 months, the rearing costs would decrease by $1,400 (≈ €1,160 or 4.3%).

They also calculated that, at an average age at first calving of 21 months, any increase in herd culling rate or in pre-weaning calf death rate caused an increased shortage of heifers to replace cull cows: up to 10-12 heifers (that is, at a pre-weaning death rate 25% and a herd culling rate of 35% – note that these are USA data!). Or, summarised in other words, each 1% of reduction in herd culling rate leads to a $1000 to $1,500 (≈ €1000) less rearing costs on a 100 cow farm.

1.2.2. Costs of dystocia in first calf heifers

A questionnaire survey by McGuirk *et al.* (2007) has shown that in the UK the economic losses of a slight dystocia could be estimated at £110 (≈ €165), while for severe dystocia this was £350 to 400 (≈ €525-600) on average.

The most important determinants of these costs were labour, followed to a lesser extent by increased number of days open, increased mortality and premature culling.

1.2.3. Comparing pasteurised, non-saleable milk feeding with commercial milk replacer feeding to calves

There are always cows with mastitis, or cows that have been treated with antibiotics for other reasons in dairy herds. What to do with this milk? Throw it away or could it be valued by feeding it to calves? A study by Godden *et al.* (2005) reported the comparison of such milk –after pasteurization – with regular milk replacer.

The milk replacer group of calves showed:
- less growth per calf per day (120 g less);
- at weaning a lower body weight (5.6 kg less);

- a higher probability for being treated (OR 4);
- a higher probability for premature death (OR 30).

Pasteurising discarded or antibiotic milk can hence be an attractive option. The authors calculated a benefit of $0.69 (= €0.57) per calf per day less costs. The method would only be valid for (large) herds which can feed at least 23 calves per day. Moreover, one has to invest in the pasteurisation equipment, and it might be more interesting to look into the udder health situation in the herd to reduce the volume of waste (mastitic) milk on the farm.

1.2.4. Estimation of economic losses due to respiratory disease in young stock

A study by VanderFels-Klerx *et al.* (2001) addressed both pneumonia in calves younger than 3 months and a seasonal outbreak of respiratory disease in maiden heifers up to 15 months of age. The components of economic losses were:
- treatment costs;
- mortality;
- premature culling;
- reduced growth;
- lowered reproductive performance; and
- milk yield loss in first lactation.

The calf pneumonia model showed an average loss of €31 per maiden heifer present (variation from €18 to €57), while the seasonal outbreak caused in average a loss of €27 per maiden heifer present (variation from €17 to €43).

In estimating the costs due to bovine respiratory disease (BRD) and its control, one should not assume that once control measures have been instituted, e.g. vaccination, that costs will be lower than if such measures had not been instituted. Mee *et al.* (1995) demonstrated that it could be economically more advantageous to treat cases of BRD as they occurred, rather than to institute prophylactic vaccination. This can occur because vaccination is an additional farm cost and is not 100% effective in BRD where mixed infection is the norm. Additional risk management plans may help to further reduce costs.

1.3. Feed efficiency in young stock rearing

Monitoring and surveillance of growth performance and the feed costs related to young stock rearing over time can provide the farmer and the veterinarian with a tool to detect pending problems and to start an in-depth analysis of a problem, if any.

When two farms are compared for the young stock growth performance, measurements often do not differ greatly (Boersema, 2006). However, when the same farms are compared for their feed costs per animal per day, the differences may be large. See for example Figure 1.2. Young stock on farm L grow faster and more gradually in the first year of life, the growth on this farm declines rapidly in the 2nd year, whereas young stock on farm S shows more growth persistency. However, if farms are compared for their feed costs per animal per day, the differences are large; farm S rears young stock for almost half of the costs as compared to farm L. In order to calculate feed costs per animal per day, one needs to know daily DM intake per (age-) group, the costs of feedstuffs per kg DM and the period during which a specific ration was fed per age-group.

An example calculation is as follows:

Feed costs per animal per day for a specific age-group is calculated as *Σ[(cost feedstuff X per kg DM × amount fed/day), (cost feedstuff Y per kg DM × amount fed/day), etc.]*

In Figure 1.2a it can be seen that on farm S, the first 4 months and last month of the rearing period are the most expensive ones. In the first 2 months expensive milk

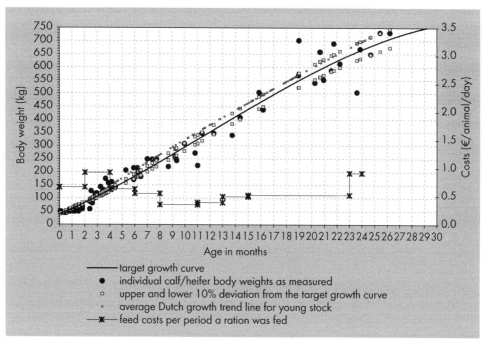

Figure 1.2a. Evaluation of growth in young stock on dairy farm S, together with data on feed costs per period a certain ration was being fed.

replacer (high protein level) is fed. During the 2 months after weaning, calves are fed a high protein calf starter, representing the most costly rearing period. From 4-8 months of age, feed costs decline as a result of the concentrates: roughage ratio getting smaller when feed intake increases. From 8-15 months feed costs gradually rise in accordance with the increase in feed intake. From 15-23 months feed intake of the same ration increases, resulting in a small rise in feed costs. Pregnant heifers are housed with dry cows 4 weeks before calving, feed costs rise because the ration of the milking cows mixed with straw was fed to this group.

Young stock feed costs on farm L presented in Figure 1.2b are much higher compared to farm S. The main difference is that an increasing amount of expensive concentrates is fed during the first 11 months of the rearing period on farm L. The concentrates:roughage ratio does not decline, resulting in high feed costs per day. In the period 11-23 a ration of very little concentrate and straw was fed on farm L., still being twice as expensive as the 'grass silage, straw and little brewers grain' – ration on farm S.

When growth performance between the two farms is compared, it can be concluded that growth in the first year is comparable. If body condition scores (BCS) are taken

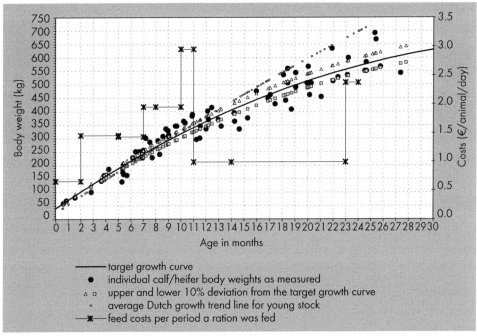

Figure 1.2b. Young stock growth charts for farm L, together with data on feed costs per period a certain ration was being fed.

into account as well, it can be observed that there is a tendency towards fattening in the age group from 6 to 12 months on farm L. High concentrate levels (resulting in high feed costs as well) may cause this tendency on farm L. The decline of growth persistency in the 2nd year of life on farm L might be caused by the poor ration fed after the 1st year of life. Note that e.g. disease losses and costs, investments in preventive programmes and labour costs are not taken into account in these calculations.

Young stock of the same breed, reared under two different farm circumstances, have developed better on farm S and under cheaper feeding conditions. Farm S has performed economically more efficient in this area.

A comparison of Figure 1.2a and 1.2b shows that one farm has much higher feed costs than the other. The reason for this difference may be in e.g. health disorders leading to poor feed intake and weight gain, poorly managed pastures, endo-parasite infections with poor feed conversion rates, and or feedstuff purchased at high costs.

1.4. Prevention and intervention: costs and benefits

Preferably, the costs and the benefits (or effects) associated with different specific intervention options are estimated before deciding which option to choose. This estimation can be done through *partial budgeting* and through *cost/benefit analysis* (or cost-effect assessment), while a *decision-tree analysis approach* can be helpful in making the proper choices when different – e.g. treatment – alternatives are presented against different prices and results (Fetrow, 1985; Cockcroft and Holmes, 2003). It is outside the scope of this text to elaborate in detail about the inputs and outputs, or methods of such estimations; we will only provide some basic outlines.

Partial budgeting is applied when a simple economic comparison of different disease control measures or of risk management measures must be made, and when the result is not depending on a time-related pattern, and does not show a high level of uncertainty. This method quantifies the economic consequences of particular changes in on-farm management, for example like in the case of applying yes or no a herd health programme, or expand an existing herd health programme with a strict anti-endoparasites component. In partial budgeting, only the cost and return elements which will most probably change because of the proposed measures are taken into consideration.

In partial budgeting, the general lay-out follows four components:
1. additional returns which will show up thanks to the proposed measures and not in the original management procedure;

2. reduced costs, showing a list of costs of the original management procedure which will be eliminated by applying the proposed measures;
3. lost returns which formerly were part of the original management procedure and which will show up as lost once the proposed measures are executed;
4. additional costs associated with the proposed measures and which were not involved in the original procedure.

The proposed intervention or prevention measures should be adopted and applied if components (1)+(2)>(3)+(4).

The other method, the *cost-benefit-analysis*, is often applied when longer term disease control programmes are at stake. By this method we can determine the profitability of such control programmes over a longer period. In this method three components are involved:
1. summing up the costs and the returns;
2. establish the discount rate (due to the longer term evaluation);
3. define the decision criterion.

The effects of the proposed measures (e.g. decrease of disease incidence) have to be transformed into economic entities. And because the various cost and benefit elements do not occur all at the same time, but most probably at various time points in the near future, we have to account for a certain (interest) change in value and price level to make costs and benefits over a period of time comparable: the discount rate (Huirne *et al.*, 2002).

The methodology largely implies the following formula:
Present Value (of a future cost or benefit) = Former Value divided by $(1+r/100)^n$

Where FV is the former (higher) value of the cost or benefit, r is the interest rate (here called discount rate; expressed in %) and n is the number of years a programme is running. Most often the inflation rate is not taken into account in detail. The discount rate is usually calculated as the difference between the current market interest rate (say 7%) and the inflation rate (say 4%) yielding the *real rate of interest* (hence 3%) also called discount rate.

The decision criterion, as third component of the method, should be chosen before starting the programme, because each of the three suggested here have their advantages and disadvantages. The three suggested criteria are:
1. Net Present Value, NPV. This NPV expresses the difference between the total present value of costs and benefits at today price level. NPV gives a scale but not a size view of the costs and benefits.

2. Costs-to-Benefits Ratio, C/B ratio. This ratio is calculated by dividing the total present value of all benefits by the total present value of all costs. This criterion does show the size (or volume) of the costs and the benefits. However, the scale of investment level is not shown by this method. The requested minimum ratio is 1.
3. Internal Rate of Return, IRR. The IRR reflects the interest rate that would make the total present value of benefits equal to the total present value of costs. In other words, the interest value that should be charged to reduce the net present value to zero. This is a simple comparison parameter and no discount rate is necessary. Point is, however, that one should recalculate at different interest rate levels until the right one is found; there are no simple formula to handle these.

Given the elements named in Table 1.2 on the one hand (Radostits, 2001a,b), and a set of parameters related to the different intervention options on the other hand, we should be able to assess both costs and benefits of a certain intervention. Intervention is defined as '*any strategy or action that should lead to recovery from disease, to premature culling, or to prevention of a certain disease in young stock*'. Such a strategy or action could comprise treatment and surgery, diagnostic testing, risk identification and risk management, or wait-and-see.

In case of neonatal disease, commonly infectious in nature, the economic loss elements are diagnostic costs, treatment costs, retarded growth (retarded growth in mild diarrhoea cases is negligible!), increased probability of death, increased probability of other subsequent disease.

An intervention A could be to wait-and-see, and intervention B could comprise antibiotics orally and/or by injection, and physiological salt infusions intravenously during 3 or 5 days in a clinic setting. In practice, only an intravenous or intra-abdominal infusion of 0.5 l of Ringer solution followed 1 to 2 hours later with a Lodevil® solution, followed by an oral re-hydration for 3 days (electrolytes) together with 2 milk meals per day at reduced volume (1 to 1½ l per day), will be feasible. See also the Annex on young stock treatment.

Intervention A does not cost anything but has an increased risk of death of the calf. Intervention B requires the veterinarian to make farm visits every day during the 3 to 5 days, and comprises the costs of treatments as well as extra labour of the farm worker. However, the intervention B has a higher probability of the calf to recover, although the recovery of daily weight gain afterwards may be reduced as compared to the calf's potential.

At the same time, one should bear in mind that in veterinary practice it is very hard to make statements on the exact probability of infection or disease, or on the exact

Table 1.2. Example of some costs and benefits of activities related to diarrhoea in neonatal calves, respiratory disease and cryptosporidiosis in older calves respectively.

Diarrhoea in neonatal calves		
	Costs	Vaccination dam
		Good hygiene disinfection
	Benefits	Less diarrhoea in neonates
		Less costs due to disease
		Better growth
		Less costs of treatment
Respiratory disease		
	Costs	Vaccination
		Good housing
		Good climate
	Benefits	Lower morbidity
		Lower mortality
		Less treatments
		Better growth
Cryptosporidiosis Coccidiosis		
	Costs	Oral prophylaxis
		Good management
		Good hygiene disinfection
	Benefits	Lower morbidity
		Lower mortality
		Better growth
		Less treatment

(After Dirksen *et al.*, 1984; Baumgartner, 1999)

success rate of a therapeutic or preventive intervention! This phenomenon is caused by, among others, the most often multi-factorial nature of disease and the unpredictable outcome of certain biological processes.

By assigning a certain cost to a given element for both the loss side and the intervention side, one is able to – at the end – consider the total costs and expected total benefits of a chosen intervention.

The basic economic health and management principle for young stock rearing should be *prevention*. Prevention (or prophylaxis) is the best investment to avoid large economic losses due to disease and/or mortality of young stock.

If we have different treatment options, or if we want to implement vaccination programmes, it is best to estimate the costs and the expected benefits, not least in order to convince the farmer to participate or not, or to assist him in his decision-making process; given the different disease histories between farms, such estimation must be done on a farm-specific basis. An example:

> Suppose that the average treatment of one diseased calf costs €50 (all-in). (Remark: €1 is set at about US $1.50)
> Suppose also that we have 30 calves at high risk for that disease on the farm.
> Suppose that if one calf gets sick, the probability that all calves get sick is 100% (respiratory virus infection for example).
> The losses would be at herd level 30×€50 = €1,500 at least (subsequent losses not included, neither is mortality). If vaccinating these calves would cost in total 30×2 times vaccination × €8 = €480 + 2×€30 extra veterinary visit costs = €540, then the benefits are clear to the farmer and it will ultimately be his choice whether or not to conduct prophylaxis.
> Note: costs of a farm visit by the veterinarian are proportionally weighed in the mentioned €50 for illustration purposes.

In the example above, the cost of losses and benefits of a preventive programme are quite straight forward, direct costs and direct benefits are clear in advance. This depends strongly on the disease one is dealing with. A disease which has major impact on the profitability of a farm business is clinical paratuberculosis (Johne's disease). Estimated losses are over €1000 per clinically affected adult cow! (excluding the losses due to culling and laboratory sampling costs when the farmer decides to lower the prevalence on his farm). Since the infection takes place in the first year of life, prevention of infection should be done during that same year. Prevention and control of paratuberculosis infection need investments in internal biosecurity, time and effort with improvements/benefits only visible many years after installing preventive measures in young stock rearing.

Moreover, the will to change and decision-making is, next to rational arguments, also based on irrational arguments, which are influenced on their turn by perceptions, impressions, emotions, attitude, motives and preferences. Humans (farmers) rather choose a status quo than changes that might lead to losses on the short term (even if these losses will be compensated for by future profits). This typical psychological phenomenon is called 'behavioural economics' (Rabin, 1998; McFadden, 1999).

Not all farmers are sensitive to the rational approach of cost-benefits calculations. Especially when preventive costs lie in the near future and disease losses and benefits of the preventive programme many years from now, it is sometimes hard to convince farmers of the existence of risks in their young stock rearing. A truly professional type of communication, addressing not only the rational arguments, but also the non-rational aspects like emotions, perceptions and attitude of the farmer can help in finding the right way of convincing and decision-making. For example, a recent Danish study showed that an important distinction between herds which had high and low calf mortality rates was whether the calf manager had a basic belief that calf mortality is a permanent crisis that has to be expected. Issues of empowerment were key to helping the calf manager to take control of and believe in his or her ability to do so (Vaarst and Sorensen, 2009).

The way and the extent to which different types of risks are managed depends on such factors as: a farmer's degree of risk aversion, the costs involved, (economic) impact of risk, the relative size of a risk, the correlation of the risk with other risks, other sources of indemnity, a farmer's perception of the nature of risk, and a farmer's income and wealth (Hardaker *et al.*, 1997; Harrington *et al.*, 1999; Barry *et al.*, 2000). Risk attitude of farmers (who are entrepreneurs these days) is in general based on positive evaluating behaviour and therefore farmers are often seen as 'risk-takers'. Farmers believe that the outcome of decisions is mostly determined by themselves, based on a feeling for their efficacy, keeping their own risk perception in mind (Bergevoet, 2005).

Entrepreneur-success is the accomplishment of multiple targets. According to Bergevoet (2005) – Dutch – farmers are mainly interested in *job satisfaction* (expressed in intrinsic values, like public image; working with animals; food safety as a primary characteristic of their business; challenges are 'chances, and no threats'). They are not completely driven by economic targets; the afore-mentioned intrinsic values of the farm are at least as important.

In general, it is assumed that farmers are willing to pay a price to reduce exposure to risk. If farmers can manage the risks on their farm at acceptable cost, they should consider themselves to be better off as a result (Arrow, 1996; Harrington *et al.*, 1999*)*. However the exact benefits of preventive programmes (e.g. in HACCP like quality risk management programmes, see Chapter 6) are often unclear to livestock producers, and substantial education is necessary to change this scepticism (Gardner, 1997).

From the text above it becomes clear that the rational approach by only taking costs and benefits into account, is sometimes not enough to convince farmers. Many irrational factors determine the risk attitude of farmers and their willingness to invest in preventive programmes. It is of major importance for the consulting veterinarian to recognise and handle the rational and or irrational drives of a farmer. Expert veterinary knowledge remains important, but above that, communication skills and sufficient psychological insight in advisory relationships have become essential too.

Farm or herd health and productivity management (FHPM, HHPM) programmes, if professionally executed, take these elements into account. They can provide the farmer with a type of coaching because on a frequent basis farm visits are made (see Chapter 2 and 3).

In general, programmes of preventive measures yield the benefits as listed in Table 1.3.

Figure 1.3 presents in a schematic way the different input areas of young stock rearing which are paramount to achieve optimal rearing results. The upper level of the pyramid can be reached when the farmer is willing to invest in a different risk-attitude and in high quality management, and to apply rather comprehensive rules and management protocols derived from e.g. farm or herd health and production management programmes. The various areas are dealt with in subsequent chapters.

The basic economic health and management principle for young stock rearing should be *prevention*. Prevention (or prophylaxis) is the best investment to avoid large economic losses due to disease and/or mortality of young stock. However, prevention is not about vaccination only; it is also about providing a good environment, an optimal comfort and high quality management. That is why a population-medicine approach to the management of young stock is proposed here.

Table 1.3. The major benefits resulting from preventive measures on dairy farms.

Major benefits of preventive programmes in dairy cattle
Reduced losses due to direct costs of diseased animals (e.g. treatment, death)
Reduced losses due to indirect disease costs (e.g. culling, discarded milk, extra labour costs)
Targets weight and age at first calving are achieved, therefore less young stock needed
Animals are able to fully develop their production and genetic potential
Increased labour satisfaction
Farms become more sustainable and more profitable

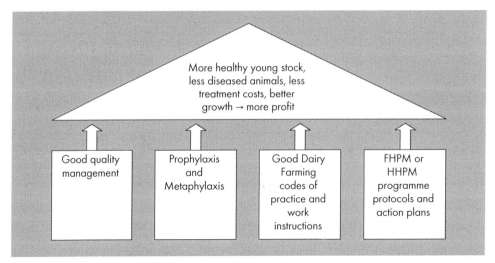

Figure 1.3. Input areas for achieving optimal end results in young stock rearing.

Chapter 2. The population-medicine approach to young stock rearing

2.1. Population-medicine, herd health and productivity management, farm health and productivity management

Population-medicine is the scientific discipline which addresses both healthy and diseased animals at population (i.e. herd, group) level, with the objective of increasing the health status of the animals and, hence, their profitability. Population-medicine is an interdisciplinary science. Quantitative epidemiology and farm economics are examples of domains involved (Noordhuizen *et al.*, 2004, 2008).

Herd health and productivity management (HHPM) programmes represent the practical execution of population medicine in the field (Brand *et al.*, 1996). Herd health and productivity management programmes have been introduced into the dairy sector from the 60's onwards, in different formats. First, there was the udder health control scheme (Bramley and Dodd, 1984), followed by herd fertility schemes in the 70's (De Kruif *et al.*, 2007). Later on, more integrated herd health and productivity management programmes have been developed and largely implemented in several dairy industries (Brand *et al.*, 1996).

Recently, the term *Farm health and productivity management* (FHPM) programmes has been introduced to better express the fact that – from a professional point of view – it is more reasonable to address not only the animals, but also their environment as well as the management regime implemented on a dairy farm.

Professionally executed FHPM or HHPM can be characterised by the features shown in Table 2.1 (after Noordhuizen *et al.*, 2008).

The *ultimate objective* of young stock rearing is to raise healthy heifers of good genetic make-up and with good body weight and conformation, which are able to produce at a high level of milk production in a durable way. This objective can be translated into technical, operational performance-related goals. These will be dealt with in subsequent chapters and paragraphs.

Farm or enterprise goals are very farm/enterprise-specific and cannot be copied from one farm to another. A farm/herd health and productivity management (FHPM/ HHPM) programme can be developed for the young stock rearing, as it has been developed for dairy cows. Providing the philosophy and the core elements of such a professional FHPM approach is the ultimate objective of this book.

Table 2.1. Overview of the 16 most important characteristics of a farm/herd health and productivity management (FHPM, HHPM) programme on dairy farms.

Mutual trust and confidence between farmer and veterinarian
Broad competencies in veterinary and zootechnical disciplines in veterinarians
Professional, functional network of other specialists to be established by the veterinarian
Proper communication skills and knowledge about behavioural economics
Minimal but necessary records are kept by the farmer and by the veterinarian
Planned farm visits with a frequency depending on herd size, stage of production cycle, and problems prevailing (at least once a month)
FHPM/HHPM provided on the basis of farmers' priorities and farming goals
The FHPM/HHPM contents are clearly defined, documented and agreed upon
The fee structure is transparent and agreed upon beforehand
FHPM/HHPM activities are well defined, organised, structured along protocols and work instructions
Routine activities comprise routine monitoring, problem analysis, and preventive measures
Cost-benefit assessments of disease losses and proposed interventions are made
Written reports on farm visits and on problem analysis are provided each time
Introduction of technical instructions for particular areas
Installation of on-site training sessions for farm workers as appropriate
A contract between farmer and veterinarian is on the basis of compulsory efforts engagement, and not on the basis of a goal-achievement-engagement

The first step in this approach is adopting a business-like attitude. This is highlighted in Paragraph 2.2.

2.2. Young stock rearing as a business: organisation, structure and lay-out

Young stock rearing can be regarded as a specific business. This business can be divided into different business units, as summarised in Figure 2.1. Each business unit has its own specifications and characteristics, its own resources, materials and methods, its own performance parameters for evaluation purposes. Moreover, each business unit has also specific disease risk periods. The latter are depicted in Figure 2.2, using the example of the colostrum period business unit.

In fact, throughout the whole young stock rearing period we can distinguish different disease risks and risk periods as is summarised in Table 2.2 and depicted schematically in Figure 2.1, 2.2 and 2.3. See also the Annexes.

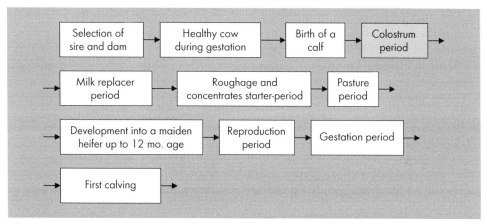

Figure 2.1. The division of the total young stock rearing business into different business units. (The shaded business unit of the Colostrum Period is further elaborated in Figure 2.2.)

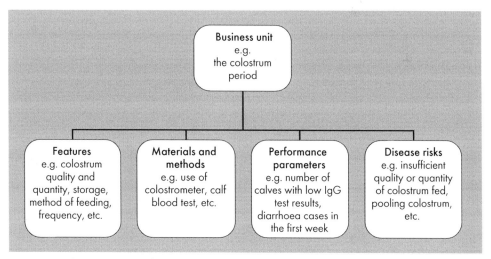

Figure 2.2. Characteristics of young stock rearing business units, using the Colostrum Business Unit as an example.

Note that the different business units can be considered as subsequent process steps in the rearing process. This implies, for example, that the business unit Colostrum Period should be properly managed in order to provide the next business unit, the Milk Replacer Period, with healthy newborn calves which drink well.

The Annexes comprise specific sheets on, for example, hygiene standards around birth, the colostrum management, and the housing of the neonatal calf.

Table 2.2. The 9 major business units of the whole young stock rearing period.

Business unit period	Age category	Features	Disease risks e.g.
Birth	0	Hygiene, assistance, separation	Oversize, mal-presentation
Colostrum	0-4 d	Quality, quantity, feeding	Inadequate colostrum management
Milk replacer	4-56 d	Quality, temperature, mixing	Poor feeding routine
Weaned	2-3 mths	Stress	Poor housing, additional stressors
Pasture	3-10 mths	Growth rate, udder development	Grass quality, endo-parasites
Housing	10-15 mths	Puberty, heat detection, vaccines	Delayed first ovulation
Breeding	15-18 mths	Heat detection, synchrony, AI	Poor heat detection, AI
Pregnancy	15-24 mths	Nutrition of dam and fetus	Abortifacients
Calving	24 mths	Hygiene, assistance	Obesity, small pelvis

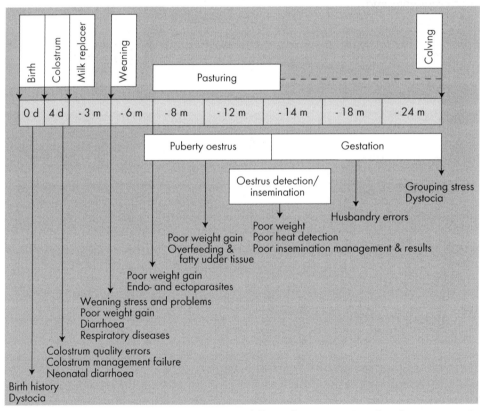

Figure 2.3. The total rearing period, divided into different business units related to age or main activity, with the main features per age category, and respective disease risks per business unit (d= day; m= month). See also the first Annexes.

2.3. Mission and farming goals

The rearing of calves and maiden heifers on a dairy farm is an investment in the future of the dairy farm. This means that the farmer (or farm manager, or contract heifer rearer) has a *general mission*, such as '*to provide, at the end of the rearing period, the milking herd with newly calved lactating heifers of better genetic make-up than the older cows, in a healthy (immune-) state and with good reproductive potential, of sufficient development and body weight, and having given birth to a healthy calf without any difficulty, while being fit for high milk yield*'.

This rearing requires many efforts during, on average, two years without any directly tangible output like milk. On the contrary, the indirect outputs are named implicitly in the forenamed mission. *Productivity* in young stock rearing comprises the issues of good health, good body development and body weight (daily weight gain) and optimal milk production potential. As a reminder, the Annexes comprise a summary of microbiology and diseases in young stock (adapted after Radostits *et al.*, 2000, 2007; Boersema, 2006; Smith, 2009).

The mission statement for rearing has to be translated into technical goals or targets for the different business units, and when needed for sections of those, as were named in Figure 2.1, 2.2 and 2.3.

The basic scheme for setting operational goals, programme execution, monitoring, making decisions and evaluating performance outcome, commonly called the *management cycle* is used. Management is pivotal in young stock rearing (Waltner-Toews *et al.*, 1986). The management cycle is presented in Figure 2.4.

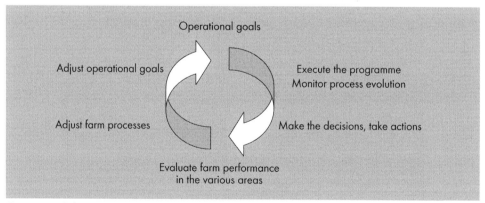

Figure 2.4. Schematic presentation of the management cycle with its respective components. (Adapted after Radostits et al., *1994; and Brand* et al., *1996)*

In every area of the young stock rearing process, this management cycle should be used to specify its respective components. These form the basic outlines of a HHPM programme.

For the weight gain targets, *target curves* for several breeds have been developed (see Brand *et al.*, 1996). Weight gain must preferably be checked once every month, twice yearly at least, using a weigh band (heart girth measurement) as necessary (see Figure 2.5 and 2.6).

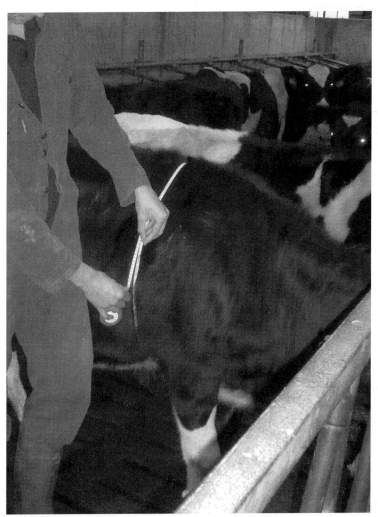

Figure 2.5. Monitoring of growth of young stock by using heart girth measuring tape. The tape indicates the estimated weight of the HF animal (Note that the cm-kg conversion on the tape is for an average HF animal, growth and development varies between farms).

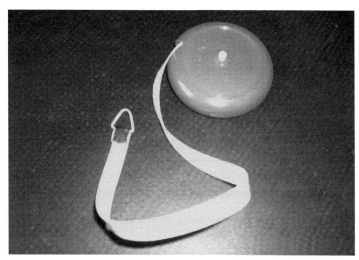

Figure 2.6. An example of a weight band for heart girth measurement of young stock. Growth performance can be assessed using a weight scale or using a weight band. The latter can easily be used in practice for heart girth measurements because centimetres can be converted into kilograms body weight. The conversion from cm to kg has to be adapted for each breed.

An example of a target growth curve together with growth data of one farm is presented in Figure 2.7 for Holstein-Friesian young stock. Note that this curve is not applicable to other cattle breeds.

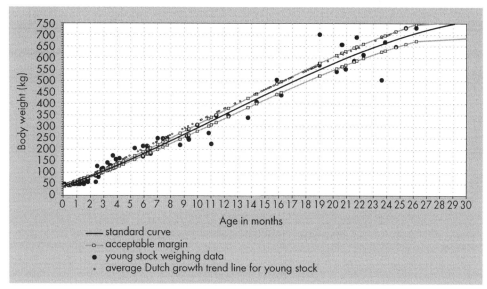

Figure 2.7. An example of a growth chart for Holstein-Friesian young stock.

For health, growth and reproduction performance during rearing, several *farm-specific target figures* have been developed (Brand *et al.*, 1996). The most important ones have been listed in Table 2.3. Performance can be evaluated each month (large herds), or each 3 or 6 months (smaller herds). For evaluation purposes, there is always a need to compare animal performance with the target figures of the farm itself (internal evaluation) and with *reference values* from a larger population than the farm itself (external evaluation). These reference values originate from a group of comparable farms or a region, or even the national level. Most of these values are either economic, biological or epidemiological in nature; that is one reason why they should be adapted to a local situation.

The target or reference values listed in Table 2.3 must be adapted to the specific farm conditions and priorities of the farmer. In order to be able to calculate these performance figures, *each animal* must be properly *identified* (e.g. animal passport or calf chart with diagnoses set and treatments given).

Moreover, appropriate *records* must be kept on the farm, e.g. in a *young stock diary* (handwritten or digital). It is advisable to mark in advance on a kind of *farm-planning-calendar* some key tasks that need to be carried out at a particular time, such as weigh check, order vaccines, dose calves, vaccinate heifers. In that young stock diary, all events of the young stock rearing are recorded. Suggested headings of such a farm diary for calves and maiden heifers are given in Table 2.4a. The event is listed in the respective column, together with date and treatment if any.

Routine monitoring (see Chapter 3) of young stock rearing can also be conducted by using the field scoring sheets which are published on the website www.vacqa-international.com. In Table 2.4b an alternative system for recording events in young stock rearing is presented. Each farm can develop its own diary, provided that the necessary components will be addressed in the most practical way.

Table 2.3. Target or reference values for exeternal evaluation of Holstein-Frisian young stock rearing performance (to be adapted to a specific situation, country or region).

		Reference values[1]
Health performance		
% of calves	born alive	> 95%
	with poor birth history (e.g. at dystocia of dam)	< 10%
	received poor colostrum quality (colostrometer)	< 15%
	with serum IgG levels < 10 g/l in colostrum period	< 15%
	with navel disorder	5%
	with diarrhoea	< 5% / < 10%
	with health problems around weaning	< 15%
	with respiratory disorders	< 5% or <10%
	with ecto-parasite infection	< 10%
	with endo-parasite infection	< 10%
	with other health problems	< 15%
	dead in the first year of age	< 5%
Reproduction performance:		
	average age at 1st detected heat	11-12 months
	average body weight at 1st detected heat	275-300 kg
	average age at 1st insemination	15 months
	average weight at 1st AI	325-350 kg
	pregnancy at 1st AI	> 60%
	pregnancy at all AI	> 95%
	heifers with > 3 AI	< 10%
	heifers with repro problems (abortion, COD)	< 5%
	heifers assisted at first calving	< 25%
	heifers with dystocia	< 10%
	heifers'calves with perinatal mortality	< 10%
	average age for final pregnancy diagnosis	17-19 months
	average age at first calving	24-25 months
	average weight at first calving	540-580 kg
	average body condition score at calving (scale 1-5)	3.0-3.25
Other management issues:		
	average age of calves at weaning	9 weeks
	availability of feeding schemes/age group	> 5 days age
	availability of strategic de-worming schemes	on indication
	presence of hygiene barrier/disinfection tub	always
	presence of a farm diary for rearing events	always
	calves with deviant weight gain	< 10%

[1] Variation in parameters occurs due to differences in husbandry systems, breeds, and/or climates (adapted after Morrow, 1986; Radostits *et al.*, 1994; Peters and Ball, 1995; Chassange *et al.*, 1999; Intervet, 2006; Collel Verdaguer, 2007; Youngquist and Threlfall, 2007; Jimeno Vinatea, 2009; www.partners-in-reproduction.com).

Table 2.4a. Example of headings in a young stock diary for the rearing period.

ID of calf	Work number	Birth date	Barn lot	Weaning weight and age	Weight + date2	Repro event	Weight at AI	Health problems	Culling date + reasons
046678	22	2-3-2009	2					Diarrhoea 11-3-2009	

Table 2.4b. An alternative example of a system for recording events in young stock rearing.

Farm diary date	Event (related to animals, management, environment)
Monday April 9	Vaccinated 33 replacement heifers in Barn 2 with BVD-vaccine *(product; dose; animal numbers)* for the first time
Tuesday April 10	Heifer 0543 coughing badly, treated with *(product; dose)* New calf pellets ordered (2,000 kg)

Chapter 3. Execution of a farm health and productivity management programme for young stock

Before starting a farm health and productivity management (FHPM) programme for young stock it is advised to make a general young stock herd inventory. This *general inventory* serves as a baseline for further activities to come (see the schematic outline in Figure 3.1). Based on the outcome of this inventory, priorities can be set (e.g. related to problem areas) for the FHPM programme, and the outcome can also be used for later comparison to monitor improvement. In Table 3.1 an example is presented of such a general inventory sheet of young stock rearing. All together, this checklist comprises 80 scoring elements, predominantly focussing on management (Waltner-Toews *et al.*, 1986). One may decide to assign an overall total score to young stock rearing (for example, overall score is good if >70 of all points were scored positive), or an area score (for example, area score is good if >70% of area points is scored positive).

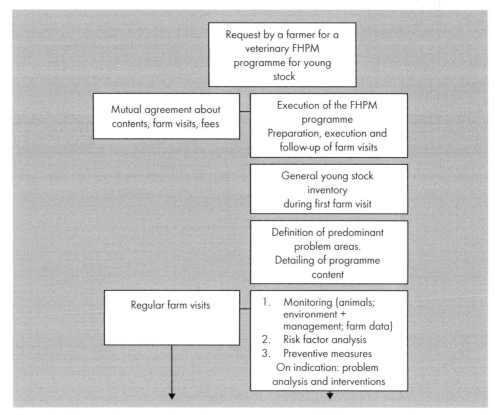

Figure 3.1. Schematic outline of activities when setting up a FHPM programme for young stock.

Table 3.1. Example of a general inventory checklist, used at the start of a FHPM programme for young stock rearing (the bold items represent the reference values for assessment).

Area + item	Scoring class 1	Scoring class 2	Scoring class 3
1. Dry cow management			
Average BCS	< 3	**3.0-3.5**	> 3.5
Supply of minerals and vitamins	**As prescribed**		Abnormal
Hygiene status	**Good**	Moderate	Poor
Udder/teats are clean	**Yes**		No
2. Calving pen			
Present	**Yes**		No
Hygiene status	**Good**	Moderate	Poor
Cleaned/disinfected before each calving	**Yes**		No
Calf separated from the dam	**Immediately**	Within 1 day	After 1 day
3. Birth management			
Navel disinfection	**Yes**		No
Calf is rubbed dry	**Yes**		No
Housing of calf	With its dam	In straw pen	**In hutch/single pen**
Cleaning/disinfection before arrival of calf	**Yes**		No
4. Colostrum period			
1st colostrum within 2 h	**Yes**	No	Left with dam
Quantity 1st colostrum	< 1 ½ l	**1 ½-2 ½ l**	> 2 ½ l
Colostrum quality	Serum IgG < 10 g/l	**Serum IgG 10-20 g/l**	**Serum IgG > 20 g/l**
1st colostrum also for 2nd feeding	**Yes**		No
2nd feeding max 8 h after 1st feeding	**Yes**		No
1st colostrum stock in freezer + identification	**Yes**		No
Colostrum intake/day	< 4 l	**4-6 l**	> 6 l
Colostrum meals / day	2	**3**	Unlimited
Colostrum hygiene	**Good**	Moderate	Poor
5. Milk replacer mixing			
Water quality checked	**Yes**		No
Mixing temperature	< 40 °C	**40-65 °C**	> 65 °C
Concentration	Unknown	± 100 g/l	**± 125 g/l**

Table 3.1. Continued.

Area + item	Scoring class 1	Scoring class 2	Scoring class 3
6. Milk replacer feeding			
Drinking temperature	Lukewarm	< 38 °C	**40-42 °C**
Use of feeding schedule	**Yes**	No	Intuitively
Check of quantity	**Yes**		No
Check drinking temp.	**Yes**		No
Litres of 1st feeding	< 1 ½ l	**1 ½-2.0 l**	> 2 l
Drinking system	**Automated**	**Teat-bucket**	Bucket
Hygiene status	**Good**	Moderate	Poor
7. Supplying cow's milk			
Feeding schedule used	**Yes**	No	Intuitively
Drinking system	**Teat-bucket**	Trough	Bucket
Dilution	Yes		**No**
Additions to milk	**Yes**		**No**
Mixing of cows milk	Yes		**No**
Antibiotic or high cell count milk supplied	Yes		**No**
Maximum quantity per Meal	< 2 l	**2-3 l**	>3 l
Fermentation applied	**Yes**		No
Pasteurization applied	**Yes**		No
8. Single pen housing			
Days in single pen	< 5 days	**5-10 days**	> 10 days
Extra water supplied	No	**Yes**	**Yes, + electrolytes**
Water supplied when	Not applicable	After the milk	**Between meals**
Concentrates supplied (kg)	No	**Calf pellets**	Cow pellets
Roughage supplied	No	Maize	**High fiber**
Hygiene status	**Good**	Moderate	Poor
9. Group housing			
Age difference/group	**< 2 weeks**	> 2 weeks	
Water supplied	**Unlimited**	Limited	No
Freshness of water	**Always fresh**	Daily	Not fresh
Concentrates supplied (kg)	**Unlimited**	Limited	
Type of concentrates	**Calf pellets**	Cow pellets	Beef cattle pallets
Frequency of concentr.	Once daily	**Twice daily**	Unlimited
Roughage supplied	Maize	Silage	**High fiber**
Freshness and quality of roughage	**Good**	Moderate	No attention

Table 3.1. Continued.

Area + item	Scoring class 1	Scoring class 2	Scoring class 3
10. Health status			
Diarrhoea	**No**	0-2 weeks	> 3 weeks
Colour of faeces at 5 weeks age	Yellow	**Brown**	Grey
Faeces consistency	**Solid**	Cake	Runny
Hair coat condition at 5 weeks age	**Shiny**	Dull	
Hair coat condition at 10 weeks age	**Shiny**	Dull	
Digestion of nutrition	**Good**	Moderate	Poor
11. Barn climate			
Air inlet	**Yes**		No
Wind breaking netting	**Yes**		No
Additional heating is available	**Yes**		No
Air ventilation	**Yes**		No
Air distribution	**Good**	Moderate	Poor
Humidity	< 60%	**60-80%**	> 80%
Air quality – NH_3	**Good**	Stench	
Infections older calves	Yes		**No**
12. Monitoring and Evaluation parameters			
Mortality rate first 3 mo	**< 5%**	> 5%	Unknown
Weight gain first 3 months of age	< 800 g/day	**800-900 g/day**	> 900 g/day
Concentrates intake at 6 weeks age	< ½ kg/day	½-1 kg/day	**> 1 kg/day**
Moment of weaning	< 6 weeks	**6-10 weeks**	> 10 weeks
Weight at 3 months	< 90 kg	90-110 kg	**> 110 kg**
Age at 1st AI	< 15 months		**≥ 15 months**
Weight at 1st AI	< 380 kg (168 cm)		**> 380 kg**
Diarrhoea prevalence	**< 5%**		> 5%
Respiratory disease prevalence	**< 10%**		> 10%
Other disease prevalence	**< 10%**		> 10%
Age at 1st calving	< 22 months	**22-24 months**	**> 24 months**
Weight after 1st calving	< 580 kg		**≥ 580 kg**

(Adapted after Brand *et al.*, 1996; Towery, 2000; Sellers, 2001; Rydell 2002; Van Trierum, 2005)

As in each professional FHPM programme, one should work according to a well-organised and structured programme. Such a programme comprises the following *core elements*:

1. *Routine monitoring* of animals, their environment and management, and farm data. It is executed during regularly planned farm visits, for example once a month, and is recorded in short, concise sheets or reports.
2. *Problem and risk factor analysis*, following a structured protocol for analyzing disorders which have a large economic impact or a high prevalence. Risk analysis and risk management are key words here. For details of such a protocol see Figure 2.4.
3. *Preventive measures* (including risk management), because it is economically much more attractive to prevent certain disorders than to cure them.

Most of these activities are executed during *farm visits* by the veterinarian. Such FHPM farm visits comprise the following 3 elements (Noordhuizen, 2004):

1. Preparation of a farm visit:
   - check the farm (problem) history, check available documents and data;
   - check current farm records – when available – beforehand.
2. Farm visit itself:
   - routine inspection of the animals;
   - routine monitoring of the animal environment;
   - checking farm-records;
   - taking samples for laboratory examination if needed;
   - plan of actions for the short and for the mid-long term;
   - start of procedure for problem analysis;
   - discussion with the farmer;
   - preliminary farm visit report 1 A4 to be written on the farm (double-copying) and to be put in the Farm Young Stock document map (see Figure 3.2).
3. Follow-up – after the farm visit:
   - consult other specialists when indicated;
   - conduct a problem analysis when indicated (if needed, additional farm visits will be done for particular areas);
   - design work instructions to implement preventive measures and to uniform operational activities;
   - prepare specific preventive measures plan;
   - design a written report (3 to 5 pages A4) and send it to the farmer, and in case there are other farm advisors, to the latter too. Try to combine the reports of different farm advisors into one concise report.

Figure 3.2. An example of a Farm Young Stock document map from the field: Veterinary Practice 'Van Stad tot Wad Dierenartsen', Loppersum, the Netherlands.

The latter three steps should also be followed by other farm advisors when they visit the farm and do inquiries; it would really help in drawing one line for the farmer and clarify the different view points.

Young stock rearing is a complex business with different business units (see Figure 2.1). Moreover, within each business unit, different features and activities can be found (see Table 2.2 and Figure 2.3). The organisation of a farm visit is to be conducted in a logical sense in order to prevent disease- spread, starting from the younger calves up to the pregnant heifers, and following each time the sequence of routine monitoring taking notes and scores.

3.1. Routine monitoring

The elements to consider in the routine monitoring are listed in Table 3.2. Note that not all elements are dealt with during each farm visit, because it is not always necessary. Body condition, as a result of e.g. breed+health/disease/surgery+nutritional status, will only change over weeks, not days; lameness may not change in weeks or months. Routine monitoring is meant to obtain early signals of potentially deviating areas and parameters, in order to prevent a high prevalence of certain diseases and/or large economic losses (*early warning principle*). And although the sensitivity of each individual parameter may be low, the fact that this activity is conducted at each farm visit and conducted by combining several parameters at the same time, will increase the sensitivity (Thrusfield *et al.*, 2001). The environmental parameters (Table 3.2) most often represent risk factors which predispose for the occurrence of diseases and disorders. If disease data and risk factor data would be available in sufficiently large numbers, the impact of such risk factors can be assessed (Thrusfield *et al.*, 2001).

The items in Table 3.2, as well as the methodology to deal with these items on the farm, have been extensively described elsewhere (e.g. Brand *et al.*, 1996; Zaaijer and Noordhuizen, 2003; Noordhuizen *et al.*, 2008; www.vetvice.com; www.vacqa-international.com).

Examples of a body condition score (BCS) chart for young stock of a dairy farm are presented in Figure 3.3a and b (Boersema, 2006). With the surveillance of the BCS data over time, one can detect potentially pending problems in weight gain, due to e.g. poor management, health disorders, poor nutrition, or poor environmental conditions. At

Table 3.2. Review of elements forming part of routine monitoring during farm visits (i.e. general diagnostic evaluation). Monitoring activity depends on farm objectives, actions planned, age-group, herd size, prevalence of disorders, management quality, time available, problems present).

Animals	Their environment	Farm data
Body condition scores	Housing (boxes, cubicles, areas)	Farm diary[1]
Rumen fill scores	Barn climate (T, RH, draught)	Milk replacer
Rumination frequency	Feed quality of roughages	instructions
Faeces quality	Feed quality of concentrates	Laboratory results
Hair coat condition	Feeding management	Post-mortem results
Skin lesions	Colostrum management	Disease records
Ecto-parasite prevalence	Milk replacer management	Growth results
Claw lesions; posture	Hygiene	Water quality[2]
Endo-parasite checks	Pasture history and- quality	AI+sire data

[1] Such records comprise e.g. colostrum data, oestrus and insemination data, pregnancy diagnosis data, semen quality checks, bull testing results.
[2] Comprising microbiological, chemical and organoleptic parameters.

the same time, the chart can be used to convince the farmer that something should be changed. In Figure 3.3a, the older young stock get over-conditioned; in Figure 3.3b, both at younger and older age there are calves over-conditioned.

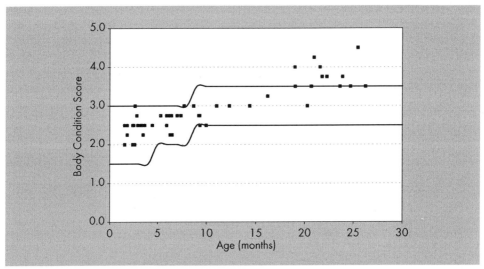

Figure 3.3a. An example of a body condition score (BCS) chart for young stock on farm S. Dots represent individual calves; the lines give the upper and lower limits of the target margin.

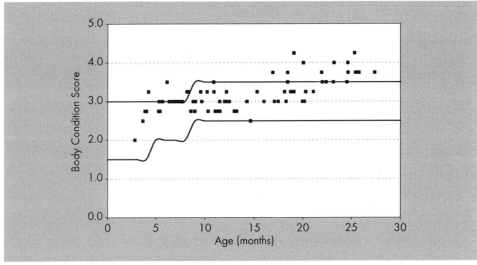

Figure 3.3b. An example of a body condition score (BCS) chart for young stock on farm L. Dots represent individual calves; the lines give the upper and lower limits of the target margin.

For monitoring the different elements in the animals, we use *scoring forms and check lists* where each element is scored from 1 (poor) to 3 (moderate) to 5 (good) at group level or herd level. The same applies for *scoring* environmental elements or farm data.

When young stock are kept in different barns and lots, it is advised to conduct the scoring for each barn or lot separately. For that purpose, different scoring forms can be designed for use in the field.

The practical objective of routine monitoring, the *general diagnostic evaluation* (of animals; environment/management; farm data) is to detect important strong points and weak points on the farm, in the herd and in farm management, in relation to animal health, animal welfare or young stock comfort, and productivity. The outcome of this monitoring is put into a *written report,* where the weak points are translated into issues for improvement. In this way, the written report comprises a *Plan of Action* for the short term and for the mid-long term. The short term (priority) advice must never exceed 3 to 5 adaptation measures in order to retain the motivation of the farmer. It is, hence, highly crucial, to install a certain timing in the execution of actions and to indicate the person responsible for a certain action. There exists a website (www.vacqa-international.com) where scoring lists for finding strong points and weak points in young stock rearing can be downloaded and – once filled in – saved for comparison with findings at a later date.

3.2. Problem and risk factor analysis

The second core component of the FHPM programme concerns the protocol of problem and risk factor analysis. A problem is either a (too) high prevalence or incidence of a certain disorder which often is not (economically) visible to the farmer at all, or a disorder with a (too) high economic impact on the farm business. This protocol comprises, in a structured way, the most relevant steps to take in a sequential order during the analysis procedure. Basically, the protocol is applicable to every kind of problem, although sometimes fine-tuning of the protocol is needed due to particular conditions of a problem. For example, a herd health problem with a nutritional basis, e.g. young stock mortality or illthrift possibly caused by a micronutrient disorder, will require a problem analysis protocol with emphasis on a whole farm micronutrient audit (Mee, 2004).We will stick to the basic model of a protocol for illustration purposes. The schematic outlines of a problem analysis protocol with its 10 steps are given in Figure 3.4.

One paramount element in this protocol regards the in-depth risk factor analysis. Table 3.3 comprises some relevant risk factors for diarrhoea in neonatal calves and for respiratory disease in calves between 2 and 7 months of age.

Table 3.3. Some risk factors for diarrhoea in neonatal calves and for respiratory disease in young calves.

Risk factors for diarrhoea	Risk factors for respiratory disease
Non-vaccinated dam	Overcrowding in the animal facility (high animal density)
Poor hygiene in the animal facilities (e.g. maternity and calf pen)	Mixing of calves from different origin
Poor hygiene at calving/birth	Poor animal transport conditions (in case of feedlot calves)
High animal density	Stress factors (housing; social; climate stress)
Failure to remove calf quickly from dam	Poor environment, poor climatic conditions (temperature, humidity, draughts)
Poor colostrum feeding management	Poor hygiene and ventilation in barns
Poor quantity of IgG per litre colostrum	No vaccination programmes applied
Poor milk replacer strategy (transition and scheme)	Poor disease detection abilities of farm workers
Poor water quality	Heat stress conditions
Environmental and other managerial stress factors (e.g. irregular feeding)	Poor management of handling bedding material (chopped straw blown into calf barn)
High animal density	

(After Radostits *et al.*, 2000 and 2007; Howard, 1993; Smith, 2009; Andrews *et al.*, 2008; Dirksen *et al.*, 1984, 2005).

The results of our risk factor analysis will provide the means to design a prevention plan of risk management. First, the different risk factors – after their identification on the farm – have to be weighed by farmer and veterinarian, and all factors ranked in descending order of estimated importance. The most important 5 to 10 risk factors are addressed in priority.

Examples of how to handle problem analysis protocols such as listed in Figure 3.4 can be found in e.g. Brand *et al.* 1996 and De Kruif *et al.* 2007 and will not be elaborated on here any further.

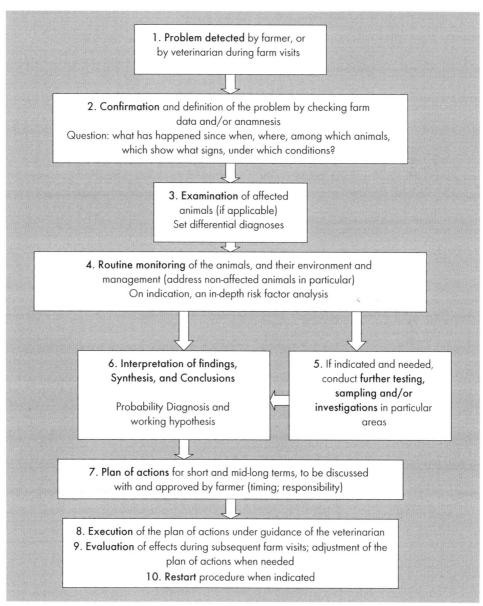

Figure 3.4. Schematic representation of the problem analysis and risk management protocol with its 10 steps.

3.3. Preventive measures (including risk management)

The third core component of the FHPM approach is the component of preventive measures. Although this may look like the smallest component in volume, it is maybe the most important one from the economic point of view of the farmer. Since preventive measures are targeted to avoid diseases entering or spreading on the farm, and hence to avoid economic losses due to disease and milk production losses, these measures must be cost-effective, highly focussed and farm-specific.

The results of our monitoring activities and an in-depth risk factor analysis will provide the means to design a prevention plan of risk management (see below).

We distinguish two different categories of preventive measures, which are either general or specific in nature:
1. General preventive measures, regarding general hygiene and farmer attitude (e.g. Good Dairy Farming code of practice, OIE, 2006; Noordhuizen *et al.*, 2008).
2. Disease-specific measures:
   a. risk identification and risk management measures, related to a specific disease or disorder (e.g. diarrhoea; respiratory disease);
   b. vaccination programmes and de-worming strategies (both not addressed in this text).

In the first category we can find the listings of very general preventive measures. Examples of such measures are given in Table 3.4 (after Noordhuizen *et al.*, 2008).

Under general preventive measures we also can find the so-called Good Dairy Farming (GDF) codes of practice, general guidelines about *how to do* certain things and how to stop executing other things. These guidelines (OIE, 2006) are focussed on creating the proper understanding, attitude and mentality in the farmer and farm workers. Some of these general guidelines are (partly) translated into *technical work instructions* and standard operational procedures for specific farming areas. For example, the guideline of Good Medicine Application may comprise a working instruction like the Young Stock Treatment Advisory Plan. The latter addresses the question '*what exactly must be done how and when*'. For details of such plans and instructions, see Noordhuizen *et al.* (2008).

In Table 3.5 we present some examples of general GDF guidelines and of technical work instructions. These guidelines and work instructions are very helpful in explaining to farmers and farm workers the relevance and responsibility of risk management. Moreover, they contribute to a quality-oriented attitude. In some cases it is necessary

Table 3.4. Examples of general preventive measures (note that the lay-out of this list facilitates its use as an on-farm checklist too).

Preventive measures	Execution		Remarks
	Date Time	Person	
Hygiene and housing			
Boxes and cubicles are cleaned			
New bedding material is added			
Exercise areas are cleaned, dry			
Biosecurity plan is operational			
Record keeping			
Sick animal records are updated			
Sick animal pen is disinfected			
Heat detection			
Heat detection records are updated			
Expected-heat-date list is used/updated			
Water, feed and feeding			
Water system is checked, running and clean			
Automatic calf feeder is cleaned and calibrated			
Feeding equipment is washed			
Other management			
Maternity pen is clean and dry before next calving			
Calves are removed from dam immediately after birth			
House the neonatal calf in a clean, dry box			

Table 3.5. Examples of Good Dairy Farming (GDF) guidelines and related work instructions in young stock rearing.

Guideline (GDF) example		Related work instruction
Good medicine application	→	Young stock treatment advisory plan
Good colostrum management	→	Preparing deep freezer colostrum
Good feeding practice	→	Preparing milk replacer rations
Good hygiene practice	→	Hygiene rules at colostrum feeding
Good pasturing practice	→	Pasture rotation schedule for different age groups of young stock

to explain the relevance of these guidelines and work instructions to farm workers and their responsibility to comply with the rules therein.

During the first inventory farm visit, the veterinarian presents his findings about the animals, the farm data, as well as the environmental issues to the farmer. Summarising the findings before the discussion is very helpful in retaining the farmer's attention. The discussion is about problems and causes, about alternative options for solutions and about losses and costs. Finally, the aforementioned issues are written down in a farm visit report. In this way a better awareness is created. If needed, a further *explanation and discussion* will be conducted when sitting around the table in the office or farm house. The findings can also be used as a starting point to conduct an in-depth analysis, for example when a problem appears to be pending or already present and is a priority area for the farmer. It is always preferable to prepare the farmer for the analysis steps that are ahead and for the options for potential solutions.

Preventive measures, oriented toward risk management, are also part of the discussion, because we have to know what would be feasible for a farmer and what not. The latter issue raises the question 'why not?'.

A golden question to be raised in that respect is: '*what are – in your opinion – the advantages and disadvantages of the action plan or preventive measures that I have proposed?*'. At each following farm visit, the same summing-up and discussion procedure is followed. In providing farm health services it is an axiom that it is '*better to progress step by step and gradually over time than by sudden, big changes*'.

Farm health service concerns a chain of events and activities; all elements should be addressed to some extent which is more effective than not to address a certain area at all. If not, the chain will breakdown somewhere and failure of the farm health service is the result.

3.4. Executing a FHPM farm visit: a summary

A farm visit in the context of a farm health and productivity management programme which is focussed on young stock rearing is built up from three components:
a. preparation of the visit;
b. executing the visit;
c. follow-up after the visit.

3.4.1. Preparation for the visit

Farm visits should be planned 1 to 3 month ahead, executed on a regular basis, on a fixed day and time of the week (e.g. Wednesday at 08:30 hrs). On a farm with 50 young stock or less, one could count with one farm visit each month, with a duration of about 1 hr. On farms with more calves, the frequency of the visits will increase and or the duration increased to about 2 hrs per visit. Each farm visit is equalled by at least one follow-up hour (e.g. for data analysis, consulting other specialists, or reporting). This facilitates the calculation of the fees for both activities, on either per head basis, per hour basis or as package deal.

The preparation of a farm visit concerns the checking of findings and advice, as mentioned in the previous farm visit report(s), the screening of latest farm information (e.g. the farm diary on young stock rearing events and/or young stock data from the farm management system on PC), and a farm calendar on which seasonal items have been listed which – based on the farm history – are relevant to address on the farm or to discuss with the farmer. Among the latter items are, for example, the timely vaccination of young stock against Bovine Respiratory Syncytial Virus (BRSV) or lungworm, the discussion of adjusting the housing facilities of maiden heifers, or the rotation scheme for young stock to be pastured.

3.4.2. Executing the farm visit

Executing the farm visit starts with implementing Good Veterinary Practice, that is, at least:
- be sure to park the car in the designated area;
- change boots and clothes;
- wash hands at the entry of the farm;
- follow other biosecurity measures when indicated at the farm, like signing in, changing equipment, providing medicines (see also Chapter 6).

The routine monitoring of animals, as named above, starts with the youngest and healthy calves and from there we go to older age groups and ill animals. The activities as named in Table 3.2 are executed, but in a selective way, that is adapted to the individual farm. Together with the routine monitoring of animals, a monitoring of the environment of the animals is conducted (detection of risk factors), and record our findings of strong points and weak points on a farm visit sheet for each pen, section or area in the rearing. All along, the farmer is accompanying the veterinarian, which facilitates the immediate discussion on findings and thoughts regarding improving certain issues. It could be indicated that samples have to be taken (blood, urine, faeces) in particular pens or animal groups for additional laboratory investigation, like serology, virology, biochemistry, bacteriology. If needed, further information and possibly samples are collected to start a problem analysis. A farm visit always ends with a discussion on all findings, conclusions and the advice or intervention proposed. This discussion is indispensable because one has to find out which activities for improvement fit well within farm management and which activities will cause more or less problems for changing management habits. Moreover, the discussion helps in training the farmer or farm workers to improve their practices.

Finally, a preliminary *farm visit report* is written (1 page A4 and preferably not more) which summarises the activities, findings as strong points and weak points, conclusions, and the proposed activities in the short run and the mid-long run for improving farm management. In case there are several farm advisors attending this farm visit, it is paramount that they stick to their own knowledge- or expertise-area. Preferably, the veterinarian takes the role of moderator of the farm visit. An example of a short-hand written farm visit report for farm XX is presented in Table 3.6. Such farm visit reports must be adapted to the individual farm.

Table 3.6. An example of a short-hand written farm visit report, with findings and actions points as well as some planning issues.

| Farm: XX | Veterinarian: YOYO | Date: April 9th 2009 |
	Strong points	Points for improvement
Animals	Body condition in all groups is okay. Rumen fill scoring results okay. Hair coat shiny in all groups. See list.	Claw health in calves from 6-12 months age seems worsened. To be checked in more detail at next visit.
Environment + management	Barn climate conditions are in order. Housing of animals is okay. Colostrum management is okay; colostrum quality is okay.	Bedding material in calves under 4 month age must be renewed once daily. Reduce the proportion of mais in ration for maiden heifers to avoid fattening.
Farm data	Weight check did not show deviations from target graph.	--
	Planning item 1	Planning item 2
Next farm visit = May 10th	Pasture plan for calves and maiden heifers to be checked and discussed.	In June, we have to check on gastrointestinal parasites (samples to be taken).
Duration of visit = 1 hr 15		

A farm visit report should be clear, short and concise, and preferably not exceed the volume of 1 page A4. If the on-farm situation indicates it, the veterinarian can design specific *work instructions* for improving staff performance in a given (problem) area. An example of such a work instruction for the colostrum management is presented in Table 3.7. Note that this work instruction is a mixture of operational and prevention issues. See also the Annexes.

3.4.3. Follow-up

The follow-up of the farm visit implies activities such as finalising the farm visit report on the computer and send it to the farmer within one week after the visit (laboratory results are sent later as an annex to this report), consulting other specialists (nutritionist; climate specialist), conducting a problem analysis according to the protocol presented in Figure 2.2, and elaborate preventive measures for the farm (e.g. work instructions or GDF guidelines). A problem analysis is always followed by a written report (3 to 5 pages A4 at the most) which must be discussed with the farmer at the next farm visit.

Note: during FHPM farm visits one should never mix up the consultancy activities with clinical activities, such as setting a diagnosis and treatment for a sick calf, or dehorning young calves. A mixing up deviates the farmer from the headlines of consultancy and instead of being focussed on advisory practice, he will start thinking to collect clinical work for each coming farm visit. Best is to make a separate farm visit for this clinical work.

The chain of farm visits is an ever changing one because of its dynamics: at each visit new issues may be put forward that need attention. However, the basic model with the 3 components always remains the same, though its content may change. This uniform basic *structure* is also needed for the farmer, so he will know where he stands. It takes time to get this structure properly adopted and implemented, but it is worth the investment. Moreover, it helps in keeping up the motivation of the farmer. The veterinarian is supposed to follow a professional way of communication (see Chapter 7) because that is the only appropriate manner to get things done or changed. Making changes in daily routines of a farmer which exist for many years is not an easy job; good communication will largely contribute to convincing of a farmer to indeed make certain changes.

For further details on the implementation of this herd health and production management approach in practice, we refer to the text books of Brand *et al.* (1996), De Kruif *et al.* (2007) and Radostitis *et al.* (1994, 2000, 2007).

Table 3.7. Example of a work instruction for colostrum management (see also Annex 10).

Colostrum management work instruction of farm X
Feeding routine:
1. For colostrum feeding, always put on the same clean coverall; feeding must be done at fixed hours of the day: 08:00 hrs; 12:00 hrs; 16:00 hrs; 20:00 hrs, and in the same feeding order.
Colostrum collection and storage:
2. Collect aseptically a minimum of 5½ l of colostrum within 1-2 hrs after birth of the calf.
3. Feed the first meal (see point 8) and chill the rest of the colostrum immediately after collection.
4. Store the colostrum in a covered bucket in a cool place, protected from pets, flies and dirt.
5. If storage is predicted to be for > 24 hrs, the surplus colostrum is to be deep frozen (-20 °C) in volumes of < 1 l to facilitate thawing.
6. Never add water, antibiotic milk or mastitis milk to colostrum.
7. Measure colostrum quality with a colostrometer (see Figure 3.5a,b and Annexes). At measuring, the colostrum must be at 20-23 °C (68-72 °F) according to manufacturer's instructions. Only good quality colostrum (SG > 1050) must be fed at first feedings.
Colostrum feeding:
8. Feed colostrum aseptically immediately after collection.
9. Always use clean teat-buckets for feeding; clean them after each meal. If a calf does not drink, use a clean disinfected stomach-tube, and insert 2 l of colostrum (see Figure 3.6a,b).
10 Never supply colostrum that contains blood or other substances.
11. Colostrum feeding schedule[1]:
2 l within 1 hr after birth (assuming an IgG level > 48 g/l colostrum);
2 l within 8 hrs after birth;
1½ l at 6-8 hrs after preceding feeding;
1½ l each subsequent meal (at 8 hrs interval);
continue colostrum feeding for 3 days.
12. Each calf must receive 100 g IgG immediately after birth, and 100 IgG within 12 hrs after birth
13. When colostrum is not available in sufficient quantity, use the (fresh or frozen) colostrum from an older, healthy cow and record this.
14. Thawing of frozen colostrum must preferably done 'au bain Marie'; not in a microwave oven, nor by heating > 50 °C. Feeding must be done at a temperature of 39 ± 2 °C.
15. Record feed refusals on the calf-chart, together with the time of the day.
16. Record also the signs of other possible deviations (behaviour; disease). In case of disease, follow the protocol of the Young Stock Treatment Advisory Plan (see Annex 4).

[1] After: Foley and Otterby, 1978; Towery, 2000; Van Trierum, 2005; Boersema, 2006.

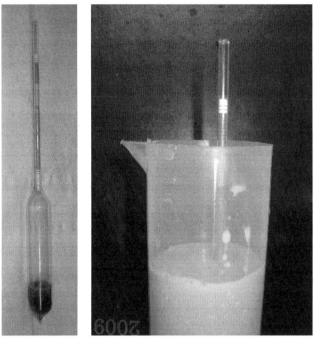

Figure 3.5a,b. Examples of a colostrometer with different colours as indicated on the scale to differentiate colostrum IgG quality.

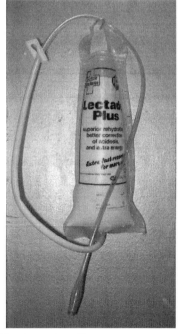

Figure 3.6a,b. Examples of an oesophagal tube for forced feeding of a calf.

Chapter 4. Applied biosecurity in young stock rearing

4.1. General principles of biosecurity

Of all cattle on a dairy farm, the calves are in general the most susceptible ones for infection and disease. Transmission of infections may occur in different ways: faecal/oral route, faecal/navel route, via colostrum or milk, nasal/saliva route, and even *in utero*. This is depicted in Table 4.1.

Given this broad spectrum of infections and their routes of transmission, it is well worthwhile to pay attention to strict hygiene measures and other preventive management measures to avoid the introduction and spread of such infectious diseases on the dairy farm, or in specific young stock areas. A core element in this management strategy regards the development and implementation of biosecurity plans.

In dairy cattle operations, the application of biosecurity plans is slowly increasing, either as a stand-alone programme, or integrated into herd health and productivity management programmes. When biosecurity functions well on pig and poultry farms, as well as on dairy farms, it must be effective in young stock rearing too. The principles will always be the same, the specifications may vary per species and per farm.

Table 4.1. Some routes of transmission of infectious agents to calves before weaning age.

	Faecal/oral	Faecal/navel	Colostrum or milk	Nasal/saliva	*In utero*
E. coli	X	X			
Clostridium	X				
Rota, Corona	X				
Cryptospor [1]	X				
Coccidia	X				
M. avium [2]	X		X		X
Salmonella	X		X	X	X
BVD virus	X		X	X	X
BLV [3]			X		X
Mycoplasma			X		
Mastitis [4]			X		

[1] *Cryptosporidium parvum*; [2] Paratuberculosis/Johne's disease; [3] Bovine leukosis virus; [4] Caused by *M. bovis*, *St. aureus* or *Str. Agalactiae*.

Biosecurity is a management instrument, a practical tool (Towery, 2000; Rosenbaum-Nielsen, 2007; Boersema and Noordhuizen, 2010).

Core elements of biosecurity are:
1. hazard and risk identification;
2. risk management; and
3. risk communication.

These are all associated with different infectious diseases. The objective of biosecurity plans is to avoid infectious diseases entering the farm and/or to spread on the farm. In designing a biosecurity plan, there are 10 main steps to take. These steps are listed in Table 4.2 (after Towery, 2000; Rosenbaum-Nielsen, 2007; Boersema and Noordhuizen, 2010).

Step 1 addresses the attitude and mentality of the farmer. It is senseless to proceed to other steps if this first step is not taken. Some general preventive measures for the dairy have already been addressed above (see Table 3.4). For young stock rearing, one may consider the general prevention rules as presented in Table 4.3.

Table 4.2. The 10 main steps in designing biosecurity plans on dairy farms.

Steps in designing biosecurity plans	Short description of each step
Step 1	Make the farm more closed (see Table 4.4) and implement general preventive measures
Step 2	Identify the most important infectious diseases and their associated risk factors
Step 3	Define farm units and sections for a better organisational structure and management; include performance evaluation
Step 4	Design and implement a farm entrance protocol for visitors and for animals
Step 5	Design and implement Good Dairy Farming guidelines (codes of practice), as well as necessary work instructions for particular areas
Step 6	Design and install an on-farm monitoring protocol in order to detect non-compliance or deviations
Step 7	Instruct all farm-workers, assign responsibilities; conduct performance evaluations each 3 or 6 months
Step 8	Develop an on-site farm-workers training programme for specific issues
Step 9	Schedule evaluation periods throughout the year to assess the critical risks in all farm units and areas defined
Step 10	Schedule team meetings at least each month and assess whether adjustment of the biosecurity plan is needed

Table 4.3. Some examples of general prevention rules for young stock rearing.

General prevention rules for young stock rearing
Strict hygiene rules have to be complied with during animal handling
Personal hygiene rules have to be complied with (coverall, boots for each calf barn or lot)
Each calf barn or lot has its own feeding equipment and utensils, which are cleaned after each meal or use
Sick animals must preferably be separated from healthy ones
Calves should be grouped on the basis of age (in large herds equal attention may be paid to body weight, development and body condition)
Preferably disposable syringes/needles should be used; if not, one needle must be used for injections in one barn or one lot of animals
For professional visitors the access road must be separated from the private access road (creation of a dirty and a clean road)
Goats must not be housed with calves in the same house
Keep calves separated from adult cattle for at least 6 to 12 months
Do not pasture calves older than 6 months in plots where adult cattle have grazed nor in plots where manure from adult cattle was applied
Feed quality (forages, concentrates, by-products) must be of optimum quality
Optimal feeding or grazing management must be applied
Quality of drinking water must be checked and be optimal; water must be always available, be clean and fresh
Veterinarian attends sick animals after healthy ones, and before moving to adult sick cattle or other species
Vermin control must be in place, as required
Unsold calves returning from the market must be refused or quarantined; bull calves born on the farm should be separated from replacement calves and housed preferably close to the area they will be picked up from by the tradesman

Next to applying general preventive measures, it is paramount to make the farm more *closed*. That means more closed against opportunities of diseases entering the farm.

The most relevant features characterising a 'closed farm' are listed in Table 4.4 (Boersema and Noordhuizen, 2010). It is obvious that these features can also be categorised under the heading of general preventive measures.

The selection of certain diseases, considered as most important on a particular dairy farm is an important step. These diseases are selected as already present on the farm, or as diseases potentially threatening the farm. For each selected disease, the

Table 4.4. Major features characterising a 'closed farm'.

Major features of a closed dairy farm
Non-essential visitors are not allowed to enter a farm, or strict hygiene measures are taken
All animal contacts (wildlife; neighbouring cattle) are avoided
Know what your neighbours are doing in health control; explain your own herd health situation
Trucks and lorries from outside are directed on a dirty road, not on a clean road
Dogs, cats, and birds are kept out of the animal buildings, while adequate rodent control is applied
Surface water is not being used as drinking water for animals
Manure from external sources does not enter the farm and is not used on farm fields neither
Own machinery is used; machines from outside must always be disinfected
The milk collection lorry follows the designated 'dirty road' to enter the farm
Cull cattle are separated from herd mates and delivered in a separate spot
Embryo transfer is not applied; visits to cattle shows are avoided
Purchased silage feed must have been harvested from fields on which cattle manure – if any – has been applied from cattle with the same health status (e.g. regarding paratuberculosis; salmonellosis); best is to record the different events in the fields throughout the year.

associated risk factors are listed and ranked in order of relevance. Table 4.5 comprises risk factors for IBR and BVD as an example (Boersema and Noordhuizen, 2010). For each individual farm one has to check which risk factors indeed are present.

Table 4.5. Examples of risk factors for two diseases, IBR and BVD.

Risk factors for IBR	Risk factors for BVD
Cattle transport trucks with mixed cattle	Running bull is used for breeding the cows
Cattle purchased from unknown health state	Cattle are purchased without testing, or no testing of calves from purchases, no vaccination applied
Infected herds in the surroundings	Pasture contacts are possible
Participation in cattle shows or expositions	Calves are housed in mixed units
Contact with other ruminants at pasture	No quarantine facility or poorly used
High stress levels in the herd may reactivate	Calving pen used for sick cattle too
Farm equipment is used with colleagues	Cattle returning from shows are accepted
Maiden heifers pastured with others	Calves are fed non-pasteurised milk
Professional visitors without hygiene rules	Persistently infected animals present
Pathogen introduction via needles, excreta	Pathogen introduction via needles, boots, etc

Risk factors which appear to be common for several selected diseases are put into a general guideline, for example a *visitors hygiene protocol*. Risk factors which are considered highly critical can be listed in a specific *work instruction* (see Paragraph 4.2).

4.2. Examples of a practical biosecurity plan

Several issues of biosecurity as named in Table 4.2 have already been addressed in earlier Tables and in the previous text; examples of other biosecurity plan components are given in the next Tables 4.6, 4.7, and 4.8.

Examples of the Risk Management procedure are presented in the Table 4.9. From Table 4.9 one can deduce that such procedures may be implemented in FHPM programmes as well, provided that these programmes are highly structured and well organised.

The art in biosecurity is to develop the most appropriate guidelines and work instructions, which does not necessarily mean many guidelines or work instructions. They must be designed when indicated or when needed because of a lack of knowledge among farm workers. In the latter case, an on-site instruction of the farm workers about how to use the work instruction properly can be warranted.

The next major component in biosecurity plans is the *monitoring*, because monitoring will tell us whether the programme is functioning well.

Table 4.6. An example of a visitors entrance protocol for farm X.

Welcome to our dairy farm! Please strictly follow the rules!
Cars and trucks must use the indicated parking area.
Visitors, please wash your hands upon arrival, change boots and clothes at the hygiene barrier.
Report your arrival by calling the telephone number named on the door; a staff member will come and guide you.
Register your name, date, reason of visit, hours of arrival and departure in the LOG.
If you need to contact animals, make sure to use our gloves which are in the hygiene barn.
Our staff will guide you over the farm.
At the end of your visit, clean boots, put the overall in the wash bin and wash your hands.
Record delivered or used medicinal products in the farm visitors LOG.
Thank you for your understanding and your visit. You helped us to maintain our herd health status, minimise infection risks and increase quality!

Table 4.7. An example of a cattle entrance protocol for farm X.

To our esteemed cattle tradesmen
New animals entering the farm must be reported to us beforehand; use the telephone or E-mail.
We do not accept any cattle delivery from a mixed animal truck load.
We expect you to wear clean boots and clothes; pass through the disinfection tub when entering.
Before delivery, the health certificates and identification papers must be handed over. No papers = no delivery.
Upon delivery, park the truck in the indicated delivery area, call the telephone number listed, and wait for someone to arrive with a lorry. You do not have access to the farm.
Put the animal in the lorry and get the proper papers signed. Our staff member will take the lorry into the farm.
Wash your hands and clean your boots before leaving.
List your name + date + time points in the LOG.
Leave the farm again by passing through the disinfection tub.
Unload animals in the designated quarantine area.
Thank you for your cooperation !

Table 4.8. Work instruction for farm workers regarding new cattle, just arrived on the farm X.

Work instruction regarding new cattle
New animals are received in a separate quarantine house; they are brought there in a specific lorry, and stay for 4 weeks.
This lorry is cleaned and disinfected after delivery.
No dogs and cats are allowed in the quarantine house.
Wash hands, change clothes and boots in the hygiene barrier in the quarantine house.
The feet of new cattle must be trimmed and disinfected in a foot bath.
New cattle are cleaned with pressure cleaner with skin-friendly detergent and afterwards disinfected.
New cattle are checked for clinical signs; upon suspicion, the vet is called or the Herd Treatment Advisory Plan is followed. If needed, samples are taken for laboratory diagnostic examination.
New cattle which are healthy (no clinical signs and or not having positive test results) and clean can enter the herd after 4 weeks.
They must be identified specifically by e.g. colour stick yellow.
After entering the herd, they must be checked for another 2-3 weeks for clinical signs.

Table 4.9. Examples of risk management procedures on a given farm. See also the Annexes.

Example protocols	Element in risk management protocol
General management	Do not purchase new cattle (nor goats, nor manure) from unknown sources or sources with lower health status.
	Make sure – once a week – that pasture contacts with other cattle or wildlife is avoided.
	Make sure that all non-farm equipment, trucks, materials are cleaned before entering the farm.
	Make sure there is each day a clean and well-organised animal health care box with medicinal products, several disposable (or at least) disinfected syringes, and some disposable (or at least new) needles available.
	Medicinal products are stored and locked away in a dark and cool place (particularly vaccines), and a record is kept.
General pen hygiene	After an animal left its pen/single box: • put the box outside in the designated cleaning area; • remove straw, manure and dirt; • clean it with high pressure water; • disinfect, rinse and let the pen or single box dry.
Calf health care	During feeding animals must be checked for sickness signs.
	Severely affected animals must be separated from herd mates and put in a special sick pen.
	Affected calves are treated according to the protocol *Young Stock Treatment Advisory Plan* (see Annex 4).
	The check result is registered on the calf-chart with date/time and name of observer.

Monitoring is the structured and formalised method to check on a routine basis the functioning of people in biosecurity, the risk management procedures and the compliance to the preventive rules that have been set in the biosecurity plan. A once a week assessment of the young stock rearing diary, the calf-charts and the performance evaluation sheets will assist in finding errors and deviations in this compliance or functioning. Sampling for laboratory examination and the laboratory results form part of the monitoring component too. In biosecurity, this type of monitoring can be easily integrated with other, operational herd health monitoring issues, as have been listed in Table 3.2.

Biosecurity plans appear to be effective (Collins and Morgan, 1992; Vonk Noordegraaf *et al.*, 1998; Groenendaal *et al.*, 2002; Groenendaal *et al.*, 2003; Weber *et al.*, 2004; Nuotio *et al.*, 2007) as has been proven for diseases such as paratuberculosis, BVD and IBR (see Figure 4.1).

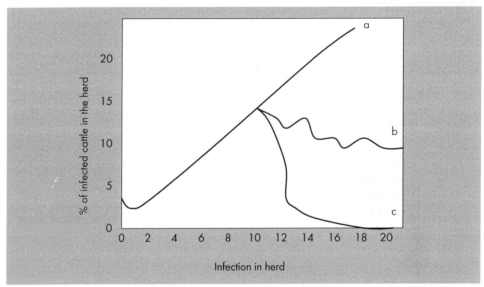

Figure 4.1. The results of an economic modelling study on the outcomes of various strategies to control paratuberculosis (Johne's disease) as adapted after Collins and Morgan (1992).
The highest infection level (a) was observed following the strategy 'wait-and-see'; the intermediate situation was achieved by the 'test-and-cull' strategy (b); the lowest level was achieved with 'general preventive measures + biosecurity measures + test-and-cull' (c).

Chapter 5. The concept of cattle comfort, applied to young stock rearing

Consumers are increasingly concerned about the quality of the food of animal origin. Quality comprises issues such as technological features and taste, but also food borne disease risks, and the way these products are being produced. Consumers want these products to be produced in an animal-friendly way. Hence, producers have to pay attention to the welfare of their animals during the on-farm production process. Moreover, welfare is an economic issue too, because it impacts animal health and productivity.

In the domain of animal welfare, there are various concepts known, some being more theoretical while others being somehow more practical in nature. The one concept best known regards the Five Freedoms described by the FAWC (1992) and by Webster (1995). However, these five freedoms appear to be somewhat anthropomorphic and not to be very practical to deal with in the daily life at a farm. Therefore, Bracke *et al.* (2002, also cited by Metz and Bracke, 2003) have developed, on the basis of the five freedoms, the so-called *12 biological needs* of the animals. These 12 biological needs are divided into 7 primary needs and 5 secondary needs (Table 5.1).

Table 5.1. The 12 Biological needs according to Bracke et al. (2001).

Primary needs
1. Feed and feed related behaviour
2. Water and drinking related behaviour
3. Resting, laying and sleeping
4. Movement (locomotion)
5. Social comfort (rank; interactions)
6. Health
7. Security (fear, flight behaviour and aggression)
Secondary needs
8. Excretion (manure and urine)
9. Thermoregulation
10. Exploration and orientation
11. Body care (grooming: comfort behaviour)
12. Reproduction

Bracke *et al.* (2001) state that animals have goal-directed-behaviour, called motivational systems. These systems are called biological needs, each having functionally related sets of behaviours or physiological responses that can be activated by certain stimuli and deactivated by a specific event or behaviour. To what extent animals are able to satisfy these biological needs, determines whether their welfare is sufficient.

For properly assessing the welfare among cows at the farm level, at least the primary needs should be translated into practical and clinical monitoring elements. Welfare disorders are, like many cattle diseases (which in fact form part of welfare), multi-factorial in nature and should be addressed following risk identification and risk analysis procedures (imbedded in HACCP-like programmes (Noordhuizen, 2005). Welfare is about adaptational processes and adaptability, which should be as optimal as possible, for both animal and farmer ('biological needs' versus 'financial needs') to ensure optimal economic profit.

The concept of *Cattle Comfort*, in its turn, is derived from the concept of the 12 biological needs, by translating the biological needs into practically observable or measurable, clinical parameters (Noordhuizen and Lievaart, 2005; Juaristi *et al.*, 2007).

Cattle Comfort comprises 4 sections, each of them with their specific parameters:
1. good quality and quantity of feed and drinking water;
2. good housing and climatic conditions;
3. good health state, especially with regard to legs and feet;
4. allowance of species-specific behaviour.

Within these 4 sections we can find the same type of parameters as we have dealt with in many previous tables in previous chapters. The various parameters of cattle comfort (see Table 5.2 for some examples) allow us to monitor during subsequent farm visits how the current state of cattle comfort is, whether trends have been positive or negative, which are strong areas and which areas need further improvement.

This approach to cattle comfort is exactly comparable to the approach of health disorders or production problems in current herd health and production management programmes, HHPM.

The previously cited website www.vacqa-international.com comprises a scoring list with different areas for scoring cattle (young stock) comfort in the field to detect strong points and weak points on a dairy farm. See also another practical publication on this issue by Juaristi *et al.* (2007).

Table 5.2. The four Cattle Comfort sections and related parameters. These parameters are fit for monitoring Calf Comfort at the farm.

Sections of Cattle (or Calf) Comfort	Parameters
Good quality and quantity of feed and water	Colostrum quality (e.g. IgG level; feeding warm, fresh, not sour...)
	Colostrum feeding practice (provided warm, satisfying amount per day, teat-bucket, 10 min. suckling time, etc.)
	Daily fresh food, with sufficient energy level per kg DM, and adequate number of animals per feeding space etc.
Good housing and climatic conditions	Dry, fresh, clean housing, no draughts
	Adequate ventilation and humidity
	Sufficient space per calf
	No contact among different age-groups
	Dry clean bedding
Good health state	Good feeding hygiene
	Good personal hygiene
	Adequate navel disinfection
	Cleaning procedures after each meal
	Optimal detection of (e.g. claw) disease signs
	Clipping the hair – at least on the back (when housed inside during winter…)
Species-specific behaviour	Social contacts (e.g. audio-visual)
	Grooming (cow brush, like tree in nature…)
	Lying with one front leg pointing straight forward

A screen example of the scoring results regarding cattle welfare and cattle comfort on a given dairy farm is presented in Figure 5.1, using the www.vacqa-international.com website information.

Cattle welfare and cattle comfort scoring can be done at the herd level, using 5 clusters of parameters (see Figure 5.1 upper part). The results are presented in a pentagram, a histogram, and – on the basis of individual scoring – a listing of good (+) remarks, moderate results (±) and issues for improvement and action (–).

On a scale from 1 to 5 (the closer to value 1, the better), on this farm the average herd score was 2.08, with areas of health, health management, management overall and

Figure 5.1. An example of scoring results for cattle welfare and cattle comfort on a given dairy farm (as taken from the www.vacqa-international.com website).

Name: João
Name of the exploitation: Pecbom - Sete Casas
Date: 2009-08-03
Total number of lactating and dry cows: 221
Average milk yield per cow per year: 8,000
Mean somatic cell count (bulk tank): 200,000 (× 1000/ml)
Mean total bacteria count (bulk tank): 10 (× 1000/ml)
Predominant breed(s) on the farm: HF
Farm type is Closed ?: Closed
Yearly culling rate of cows: 10
Is calving pattern spread over the year: Yes
Number of F.T.E. working on farm: 8
The sires/bulls predominantly used on this farm show proper traits regarding calving ease, udder health, claw/leg health, durability?: No

	Herd average
Health & Health Management	2.49
Behavioural Aspects	1.29
Housing, Equipment & Climate	2.43
Feed & Feeding Management	1.45
Management Issues	2.00
Weighted average	2.08

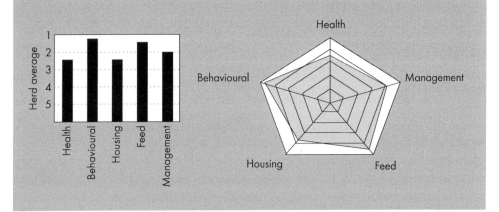

Individual parameter scores	+	±	–
Health and health management			
Yearly prevalence of acidosis/SARA	X		
Yearly prevalence of ketosis/severe NEB	X		
Herd locomotion scores are okay	X		
Yearly prevalence of lameness	X		
Yearly prevalence of hock problems	X		
Herd claw trimming and checks at least 2x/yr	X		
Formalin footbathing done properly			X
Prevalence of skin lesions/swellings		X	
Yearly prevalence of clinical mastitis			X
Yearly prevalence of infectious diseases		X	
Good medicine application code of practice operational	X		
Rumen fill scores on fresh cows okay	X		
Herd health certificates are present	X		
Behavioural aspects			
General behaviour of cows is quiet	X		
Cows can be easily approached, no fear	X		
Man-animal interactions are correct	X		
Cows behave quiet in the waiting area	X		
What is the percentage of idle standing cows?	X		
Cows are quiet during milking process	X		
Cows lay not down in exercise areas	X		
Housing, equipment and climate			
Feed rack type, access, position and size okay	X		
Cubicle sizes or deep litter are okay	X		
Cubicle bedding material is adequate		X	
Exercise areas are dry, clean, non-slippery; and of proper size			X
Floor maintenance is okay (condition slats and slots is okay and rubber mat topping is applied)	X		
Separate calving and sick cow pens	X		
Light regimens day/night/parlour are okay	X		
Ratio of # cows to # cubicles is okay		X	
Ratio of # cows to # feeding places okay		X	
Cow barn position, side walls ventilation OK	X		
Ventilation checks conducted are okay	X		
No moulds, no dirt or spinrags in cow barn	X		
Cows do not suffer from heat stress		X	
Is climate actively controlled by farmer	X		

Individual parameter scores (Continued)	+	±	−
Feed & Feeding Management			
Feed is available at feed rack	X		
Feed is of observed/tested good quality	X		
Drinking water of tested, good quality	X		
Sufficient drinking places present	X		
No moulds or soil present in silages	X		
Forages provide good structure in ration	X		
TMR is fed to lactating cows properly	X		
Particle length in TMR is correct	X		
Pasturing is being applied		X	
Feed changes are rare and gradual	X		
Management Issues			
Hygiene scores on cows, barn, parlour and feed bunk okay	X		
Herd Treatment Advisory Plan is applied	X		
Udder Health Control scheme operational	X		
Herd Health & Production Management	X		
Yearly incidence of dystocia + C-sections		X	
Endo-/Ecto-parasite control scheme operational	X		
Cattle comfort tools provided (eg brushes)		X	
Notes or Remarks			

housing indicated for overall improvement. On the basis of these outcomes, an action plan can be designed and discussed with the farmer. Once executed, a follow-up should be done to evaluate the effects of actions taken and interventions conducted.

There are many items in this scoring system which are quite comparable to the issues that have been dealt with in the farm health and production management (FHPM) approach in previous chapters. This makes it particularly interesting to integrate the FHPM approach with the cattle welfare and cattle comfort approach as presented here.

Chapter 6. Quality risk management programmes for young stock rearing

(based on the HACCP-principles)

Entrepreneur-like dairy farmers, especially those with hired labour, may wish to organise the young stock rearing on their farm in a much more formal and structured way than provided through FHPM alone. In those situations we suggest to use the *Quality Risk Management* type of approach based on the principles of hazard analysis and critical control points, (HACCP), (CAC, 2003; FAO, 1998). As has been pointed out elsewhere (Noordhuizen and Welpelo, 1996; Cullor, 1998; Lievaart *et al.*, 2005; Noordhuizen *et al.*, 2008; Boersema *et al.*, 2008), the HACCP-like approach on dairy farms provide farmers with the means to structure, formalise and organise their activities in such a way that animal health and welfare, as well as public health and food safety are safeguarded in the best possible way. The HACCP principles (Pierson, 1995, Cullor, 1998; Noordhuizen *et al.*, 2008) allow the identification of the most important hazards (diseases) and associated risks on a dairy farm, and to define critical control points for the control of the production process on the farm by developing production process diagrams, monitoring schemes and prevention and correction measures. The rearing of young stock on a dairy farm can also be regarded as a production process.

To develop a quality risk management programme on the basis of the seven HACCP principles we follow the 12 steps as presented in Table 6.1. These 12 steps will, by and large, be followed throughout the following text. Steps 1 to 5 can be considered as the preparatory phase in the development.

The quality risk management team (step 1), hereafter named as *team*, comprises the farm manager and the veterinarian, as well as the nutritionist. If needed or indicated by prevalent problems or the farming area of concern, one may add other specialists to the team, like economists, an AI specialist, an extension specialist, a public health specialist. This *team* will develop the programme and its components.

The step 2 and 3 issues have been dealt with in the paragraphs on FHPM and are, therefore, not further elaborated here.

Step 4 on the flow diagrams is a very important one because these flow diagrams will provide a clear overview and greatly assist in the discussion within the *team*, of the *team* with the farm workers, and with the external professional advisors. Flow diagrams must be validated on-site with all people involved.

Table 6.1. The 12 steps for developing and implementing a HACCP-like quality risk management programme for young stock rearing (adapted from Noordhuizen et al., 2008).

Developmental step	Short description
Step 1	Assemble a multidisciplinary, facility-based *team*
Step 2	Describe the final product (heifer) and the way it is introduced into the herd of lactating cattle (processing requirements)
Step 3	Identify the way these heifers are handled and or identify the targeted receiver of the heifers
Step 4	Develop flow diagrams of the production process of rearing young stock from birth to first calving heifers
Step 5	Verify the flow diagrams on-site on correctness with *team* members and farm workers
Step 6 (= principle 1)	Identify the most important hazards (diseases, disorders) on/for the farm, their associated risk factors. Conduct risk weighting, prioritise risks
Step 7 (= principle 2)	Identify critical control points (CCP) in the production process, required to reduce or eliminate hazards/risks. Identify points of particular attention (POPA) which can help in reducing/eliminating hazards and risks
Step 8 (= principle 3)	Establish critical limits and norms for CCP, and target figures for POPA. Deviations should trigger the implementation of corrective measures
Step 9 (= principle 4)	Establish an on-farm monitoring scheme and its requirements regarding CCP and POPA. The results of monitoring are used to adjust procedures on the farm and to maintain control of the process. Herd performance can also be assessed through this monitoring procedure
Step 10 (= principle 5)	Determine corrective measures beforehand. Implement when monitoring points to deviations
Step 11 (= principle 6)	Establish sound record-keeping procedures which document that the programme is functioning and effective
Step 12 (= principle 7)	Establish internal validation and external auditing procedures to verify that the programme is working correctly

Figure 6.1 provides the overall flow diagram of young stock rearing on a dairy farm, from birth to first lactation. We define 4 main periods in rearing (see also the Annexes). The different activities are defined and the process steps designed in their logical order. Although the basic blueprint for flow diagrams shows a lay-out which is universal for all farms, the ultimate flow diagrams are highly farm-specific. All farm workers should be informed and should be familiar with reading these flow diagrams properly. Their practical experience may be useful to adapt the flow diagrams to

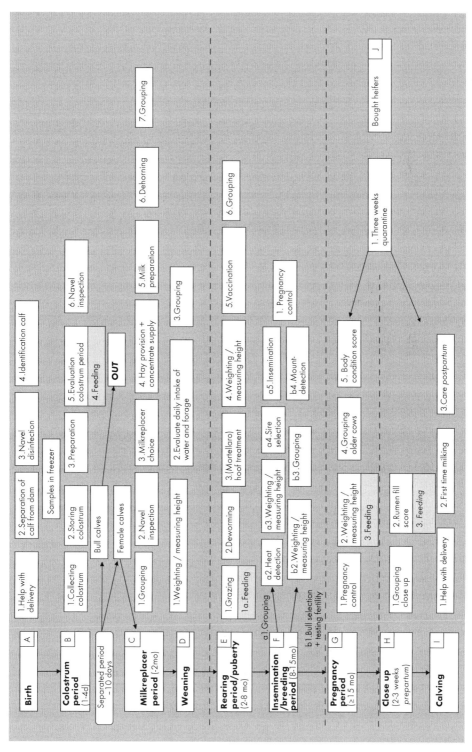

Figure 6.1. Flow diagram of the production process of young stock rearing: from birth to first lactation heifer (after Boersema, 2006).

the conditions on-site (step 5). The flow diagrams will also contribute to a better understanding of the individual worker contribution to the total process of young stock rearing. If someone recognises his own place in the total context, this will greatly improve dedication and motivation to fulfil tasks in the best possible way.

In step 6 we identify within the *team* the most important diseases and disorders (hazards) which are either prevalent on the farm or which the farmer does not like to have on the farm. These may be diseases such as Bovine Respiratory Complex, BVD and IBR, or diarrhoea, mastitis and gastro-intestinal parasite infections. Among the disorders may be distinguished e.g. claw lesions, welfare or cattle comfort disorders. At the same time, lists of associated generic risk factors are set up (see also the Tables 2.3 and 2.4). With these generic risk factors one can check which ones are prevailing on the farm and which are not, exactly like has been described in Chapter 6 on biosecurity.

In the Annexes 1, 2 and 3 there are several tables with the most important diseases and disorders (and the type of hazard, being animal health or welfare, or public health/ food safety hazards), as well as their associated, generic risk factors, distinguished per period of rearing, as well as the high risk periods of most hazards. With the generic risk factor lists, we have to detect those risk factors which are prevailing on the dairy farm, exactly like we have done in Chapter 4 on biosecurity.

Next, the *team* will determine which risk factors are more important than others; one can use the formula 'Probability × Impact' for that purpose (Heuchel *et al.*, 1999). Then, the risk factors are put in diminishing order of relevance, and the most important 10 or 20 are put in the Hazards and Risks Table. An example can be found in Table 6.2 (Boersema *et al.*, 2008).

In step 7, we have to define the critical control points (CCP) and the points of particular attention (POPA). Critical control points have to meet several formal criteria to be considered as such. These criteria are that they must:
- be observable and/or measurable;
- be associated with the disease/disorder of concern;
- have a norm (standard) and upper and lower tolerance limits;
- be crucial for process (step) control;
- be accompanied by corrective measures, which in case of lost control
- be able to fully restore process control.

In young stock rearing there is always the issue of biological variation (biological variation between animals; variation in diagnostic tests with false-positives and false-negatives). Due to these issues, not all CCP-criteria can easily be met. Especially the

Table 6.2. An example of a Hazards and Risks table from a HACCP-based quality risk management programme with critical control points (CCP), and points of particular attention (POPA) and the monitoring scheme (adapted after Boersema et al., 2008).

Hazards	Main risk factor	CCP[1] POPA[2]	Monitoring	Refers to work instruction
Diarrhoea (in neonate or older calf)	Insufficient colostrum intake	**· Monitor prevalence** **· IgG level serum in** **calves 2-5 days old**	· Monthly data evaluation · Monthly IgG level serum check or · IgG level colostrum check (each calf)	'Feeding newborn calves'
Diarrhoea in neonate	Poor hygiene	**· Monitor prevalence** *· Pens are cleaned prior to re-housing (according to 'Golden cleaning-standard')*	· Monthly data evaluation	'Hygiene around newborn calves'
Poor weight gain 12-24 weeks age	Insufficient feed/protein intake	**· less than 5 % of animals below target weight**	· Young stock measurements 2x per year	'Young stock measurements and BCS'
Poor pregnancy rate in maiden heifers	Poor weight gain 12-24 weeks (see above)	*· # insemination/animal*	· Monthly data evaluation	'Feeding young stock of 12-24 months age'

[1] CCP indicated in bold; [2] POPA indicated in italic.

third and the last criterion named above are hard to meet on livestock farms, and therefore, we can add the points of particular attention (POPAs) which are crucial in process control but which do not meet all formal CCP-criteria (Noordhuizen *et al.*, 2008). Some examples of CCP and POPA at farm level are given in the Hazards and Risks Table 6.2. The POPAs do not have a standard value; they have target figures set, exactly like in FHPM programmes (see Chapters 2 and 3).

The standards and upper/lower tolerance limits for CCP, and the target figures for POPA need to be defined in step 8. The former may originate from agricultural handbooks, the latter have to be fixed by the *team*.

Once, the CCPs and POPAs have been determined, we have to define the methods to monitor these points (step 9). An example is already presented in the Hazards and Risks Table 6.2. Included must be: method of monitoring, frequency of monitoring, where to record the results of monitoring, the person(s) responsible for the monitoring. Usually, a special monitoring table is used on the dairy farm for that purpose. Laboratory examinations (including *post-mortems*) and their results are also recorded in the monitoring table (for its headings, see Table 6.3).

Corrective measures are developed for each critical control point, and preferably for the relevant points of particular attention. This is a great difference with current HHPM programmes: the sets of corrective measures are now well-described beforehand in a Farm Health and Productivity Plan. They are classified in such a way that every farm worker can consult these sets of measures because they are in written format and or in the computer on the farm.

The corrective measures for the POPAs are highly comparable to the ones we have dealt with in the Chapters 2 and 3 on FHPM and Chapter 6 on biosecurity. However,

Table 6.3. Headings of a monitoring table, focussed on deviations observed in the process.

Nr of animal	Date	Deviation observed	Responsable person	Action taken	Therapy	Withdrawal period for milk/meat	Lab results	Action taken	Evaluation

it appears to be very practical to design *work instructions* for specific areas on the farm which appear to be in need of particular attention from the farm workers. A major example is the Young Stock Treatment Advisory Plan (see Annexes), where next to the farm diagnosis, the treatment options are listed beforehand, as well as the dosage, route of administration, withdrawal time for meat, and further instructions are listed.

Other examples, appearing in the Annexes, are: Good Colostrum Feeding practice; Good Hygiene practice in Neonatal Calf Care; Good Hygiene practice in older calves and heifers; Good Cleaning and Disinfection; Operational Management sheet on Diarrhoea.

Like in other programmes such as FHPM, we need specific, *formal records* in the quality risk management programme if it is based on the HACCP principles (step 11). Some of these records have already been mentioned, such as the Production Process Flow diagrams, the Hazards and Risk table, the Monitoring table. Others have been addressed in other chapters of this book: good dairy farming codes of practice (GDF guidelines); biosecurity plans for controlling infectious diseases on the farm.

The optimum way to handle these records and documents is to integrate them into one *QRM-Handbook* for the farm. This on-farm handbook will comprise the following chapters:
1. Introduction
    a. mission statement of the farm operation
    b. names, addresses, telephone numbers, E-mail addresses of all professional advisors and service providers
    c. procedures for emergencies
2. A scheme with the 12 steps to develop the HACCP-based quality risk management programme, the team members on that particular farm of the current year, the meeting scheme for the coming year (dates, hours, place, people), handbook revision meeting
3. The developed production process flow diagrams
4. The hazards and risks tables
5. The monitoring schemes, the operational management sheets (work instructions)
6. GDF-guidelines and deduced work instructions
7. The biosecurity plans in full
8. Other documents deemed necessary for the functioning of the programme (e.g. logs for recording corrections executed and their respective effects)
9. Training – and coaching – schemes of farm manager and or farm workers
10. Items resorting under the internal validation and external verification

The availability of GDF-guidelines and work instructions is not a formal part of the quality risk management programme, but they are indispensable for a good functioning of that programme and must therefore be integrated into the quality risk management programme. Some authors state that a quality risk management programme even starts with the adoption and execution of these GDF-guidelines in order to focus attention of the farm workers and build a stronger mentality (Noordhuizen *et al.*, 2008).

The integrated programme of HACCP-based quality risk management with GDF-guidelines and work instructions as well as biosecurity plans is in fact a Total Quality Management programme, TQM (Noordhuizen and Welpelo, 1996).

The *training and coaching* component as listed as a separate chapter in the QRM-Handbook is a highly essential one. It is useless to only address the farm manager with the QRM issues, and neglect the farm workers. The latter must be involved as much as possible: in discussions with the *team,* in communicating decisions that have been made, in executing the different components of the programme. Once they have a certain level of responsibility their motivation, engagement and performance will further improve. When deficiencies among farm workers have been reported by them and or detected by the *team* in certain areas, specific on-site short training courses can be developed and given. This investment in human resource empowerment aids in a better functioning of the programme and hence to a better functioning of the farm as a whole.

In step 12, we deal with internal validation of the programme and external auditing of the programme functioning (verification). The latter part is outside of the scope of this book and will not be addressed here in much detail. It should be sufficient to say that if a farmer wants his on-farm production process to be certified (possibly driven by the supplier demands in the food chain), he will need to address this step. This could be true in particular for farmers who only rear young stock from and for other dairy farmers. At the time of publishing this book there are no formal institutions present to conduct this type of audits on (young stock sections of) dairy farms, to the authors' knowledge.

The internal validation, however, is within the scope of our farm health activities. It regards, among others, the evaluation of the performances of the young stock herd in its different groups by means of our preset parameters and targets (see Chapters 2 and 3). Again the difference between HACCP-based programmes and FHPM is that in the former we have to conduct such evaluations at preset times in the year, for example twice yearly (e.g. in early spring and in late autumn), while this is at free choice in the latter. The internal validation is meant to assess the efficacy of the farm quality risk management programme. Next to the list of parameters and target figures, we address the hazards and risks table, and the monitoring scheme and discuss within the *team* what has gone right and what has gone wrong, and – in the latter case – how it has been counteracted. When deemed necessary, the *team* adjusts specific components of the programme, or alter the lay-out of certain tables and work sheets. The objective of this evaluation is to improve the programme, its functioning and efficacy.

It is advised to involve the farm workers in the ultimate presentation of the validation results, and give them the credit and compliments when the programme has been running smoothly and give constructive criticism when the programme has failed to a certain extent.

Again it must be stressed that it is better to proceed gradually and in all domains implicated always to a certain extent than too rapidly or while neglecting one or more domains. In the latter situations, the farmer and co-workers get de-motivated easily and the programme fails.

Chapter 7. Discussion and conclusions

Dairy husbandry has dramatically changed over the last decades. Mixed farming has evolved into mono-species farming like dairy cattle alone. At the same time, up-scaling took place from small to larger farm sizes; from family-run operations of up to 150 cows to large dairy enterprises of more than 1000 cows.

Due to a higher level of technology (milking machines and milking robots), new feed technologies (total mixed rations; movable feed racks; concentrates dispensers), input of sires with high genetic merit, improved artificial insemination and embryo transfer procedures and an increase of the number of cows per man and per hectare, labour productivity has increased. Consequences of this intensification have been the decrease in time spent per animal and the occurrence of so-called production diseases or management-diseases (calf diarrhoea, respiratory disorders, claw disorders, metabolic and reproductive disorders).

In different countries, *veterinary herd health and production management (HHPM) advisory services* for the different farming areas have been implemented by bovine practitioners and farmers to better deal with these diseases and disorders (Brand *et al.*, 1996). The herd was the unit of interest. However, predominantly lactating and dry cows have been involved in such programmes, while young stock became the forgotten population. As a result, the early detection of disease in young stock has become much more difficult and treatment often comes late, while – on the contrary – it would be more sensible to invest in disease prevention or health promotion than diagnosing and treating sick animals alone. Therefore, a more holistic approach has been proposed, where animals, their environment and the farm management are addressed at the same time, the veterinary *farm health and productivity management (FHPM)* programmes (Noordhuizen *et al.*, 2008). These programmes are much more professional in nature, and better structured.

In general, dairy farms do well in the areas of handling animals, feedstuffs, cleaning and disinfection procedures, maintenance and surveillance, as well as pasturing and pasture exploitation. However, there are other issues that need attention such as drinking water quality, animal treatment procedures, hygiene, management and prevention, controlling barn climate conditions. These are all managerial aspects. They are caused by a lack of observational skills, lack of time or knowledge and or awareness about e.g. risks, lack of implementing certain measures, inconsistencies in managerial procedures, lack of self-criticism, changes in attitude or perception, unawareness about losses involved (IKC, 1994).

Veterinary farm health and productivity management programmes (FHPM), including the *biosecurity plans* (the latter may imply a self-evaluation or farmer's own statements, as listed in Table 7.1), provide a platform for proper operational decision-making and prevention by focusing attention on the strong points and weak points in farm management and herd health of young stock; they provide the farmer with a structure and a veterinary coaching to risk identification and risk management, and hence a preventive approach.

Several chapters in this book have highlighted this type of approach. The FHPM approach is economically sound and justified; it is the best way to optimise performance

Table 7.1. Example of a short checklist of farmer statements regarding the biosecurity plans in young stock management (from: DQA, 2000 in: Noordhuizen et al., 2008).

Veterinarian's remarks on farmer statements	YES or NO	Best management practices checklist
		I or the calf raiser never allow replacement heifers on pastures where manure from mature animals has been spread (micro-organisms may live in soil for up to 1 year)
		All calves are fed colostrum from cows that have been tested clean of infectious diseases (colostrum is not mixed; each calf receives milk from a single cow)
		I avoid adding whole milk to extend purchased milk replacer (milk replacer is considered free of infectious pathogens)
		Replacement heifers are kept separate from other animals for at least 6 months (kept away from manure of adult animals)
		I ensure that there is no stagnant water in the replacement heifers pens
		Replacement heifers have their own separate source of water
		Calves are separated from their dams immediately after birth (no sucking, no searching for teats allowed)
		The calving area is always clean and disinfected
		I consult with my veterinarian (at least) annually about calf care
		Calves are permanently identified before any grouping
		All replacement calves are given a reticulo-ruminal magnet at initial breeding
		I have a strategic de-worming practice plan in place
		I routinely dip the entire navel of newborn calves with strong iodine or other disinfectant

in young stock. Furthermore, the development and application of good dairy farming guidelines (GDF) and the practical work instructions derived from those for dairy farms (FAO, 2004; Cannas da Silva *et al.*, 2006) provide a good basis for expanding FHPM to a more quality-oriented way of dairy farming. These guidelines and work instructions can be easily integrated with operational FHPM programmes, the best examples being the integration of biosecurity plans for preventing infectious diseases from entering in young stock, and the Young stock Treatment Advisory Plan. Other examples have been presented in this book.

Quality failure costs are either systems costs or true (management) failure costs, or a combination of both. Quality failure costs represent missed income, due to e.g. disease costs, costs of barn renovation, wasted labour (and often unknown losses which may be hard to identify). Such costs have been estimated at €150 to 250 per average cow present in the herd; an improvement of €100 per average cow present must be achievable. In young stock rearing, although hardly addressed in animal health economics literature, it should be feasible to reduce rearing costs and disease losses with, on average, €50 per average young animal present.

The adoption of GDF guidelines and work instructions by the farmer and farm workers would be a sound foundation for installing Quality Risk Management programmes (QRM) on the basis of the seven HACCP principles, because they adapt the mentality and attitude of people. The application of the HACCP-concept to animal health on dairy farms is a logical move because HACCP first of all focuses on microbiological hazards and risks as can be found in public health and animal health. Moreover, it focuses on hazards of a different kind like chemical and physical contamination of products, but also on disorders of another type like welfare disorders (Noordhuizen *et al.*, 2008).

A QRM programme based on the HACCP-concept can be best designed and implemented when beforehand the appropriate foundation has been laid (see above).

This foundation comprises:
1. a professionally executed veterinary FHPM;
2. the development and implementation of Good Dairy Farming codes of practice and work instructions; or
3. preferably, both.

The most important characteristics in HACCP are that *structure* is provided, that on-farm *organisation* and *planning* are installed, and that the various *steps and procedures are much more formalised* than in, for example, veterinary herd health and productivity management programmes.

> *'Applying HACCP may seem unwieldy, but it is nothing more than what a truly good farmer would do anyway'.*
>
> *(Ryan, 1997)*

The starting point for the development of a QRM programme on the basis of the HACCP-concept is either a complaint from the farmer about the performance of his herd, a deviation in herd performance as detected by the veterinarian in his FHPM, or the wish of the dairy farmer to be supported routinely in his quality control activities. This QRM approach is very much similar to the FHPM approach.

In both situations, an assessment of strong and weak points regarding animals and their environment, and the management is warranted. This provides the basic elements for both operational veterinary herd health and production management and quality risk management at a more tactical level. For herd health and production management it provides clear-cut issues for operational control and for intervention, for the quality risk management it represents the first analysis of hazards and associated risks. These features already show that activities in the areas of good dairy farming, herd health and production management, and quality risk management can and should be integrated as much as possible.

The ultimate merger between FHPM and HACCP-based QRM, has as consequence that the execution of the FHPM has to become a much more formal one, better organised and structured, exactly like the HACCP-based QRM-programme is designed. Overall, the integration of both approaches makes the veterinary service to the (dairy) farm more professional, more efficient and more beneficial for both the dairy farmer and the veterinarian.

> *HACCP-based applications in QRM programmes are not the panacea for solving all disease and productivity hazards.*

The formalisation, organisation and structuring issues are elementary components of the HACCP concept, and are required by third parties to ultimately get a reliable insight into the functioning of the HACCP-based QRM programme on the dairy farm. It should be clear to the farmer, his co-workers and the veterinarian that it is far better to apply all components of the HACCP-like programme to some extent (preferably the largest extent) instead of just applying some components! The latter will undoubtedly result in a zero-efficacy, because too many and or paramount domains remain untouched (Noordhuizen *et al.*, 2008).

It can be stated that veterinarians have a role to play in these areas because they are most strategically positioned in the field and have the best basic skills and knowledge to conduct such programmes. In many countries, veterinary herd health and production management programmes are operational; in other countries veterinarians largely contribute to dairy farm success by designing and implementing biosecurity plans. The veterinary-zootechnical background of the veterinarian must be thorough and of high quality; a thorough training in individual animal medicine largely contributes to his standing. In some instances veterinarians contribute to the development of good dairy farming guidelines. But before being able to integrate all forenamed components into an integrated Quality Risk Management programme based on the HACCP-concept and principles, it is required that he adopts and acquires new skills and knowledge before being able to fully function as a quality coach-consultant.

The developmental process for a veterinary practice, evolving from a purely curative practice to a practice where curative work is coupled to advisory activities as illustrated here takes time and investment from the participating veterinarians. Each veterinary practice has to define for itself, which goals should be reached, when and at what pace (Cannas da Silva *et al.*, 2006).

Irrespective of the fact whether a FHPM for young stock, or a biosecurity plan, good dairy farming codes of practice and work instructions, or QRM are introduced on a dairy farm, in order to be successful there needs to be an appropriate, professional communication between farmer and veterinarian (Kleen, 2008). Veterinarians with good communication skills tend to be more successful and less prone to stress (Brandt and Bateman, 2006).

While in a curative practice the communication type is rather of the '*you should do this*' or '*I will do that*' kind of approach, in the advisory practice it will rather be a '*you have a problem in young stock, I confirm that, and now let us work together and try to solve the problem*' approach.

Professional communication can be divided into a verbal one and a non-verbal one. The former is the rather technical content of a message or advice, while the latter comprises elements like outer appearance, gestures, facial expression, body language, dominance, voice. Like in decision-making, there is a certain rationality and non-rationality. Perceptions, impressions, emotions, motivations and preferences play a substantial role in the latter area. It has been stated that the technical content of a written or oral message only accounts for about 30% of the effect of that message, while the other 70% is accounted for by the non-rational elements. Six main factors influencing the interaction between discussion partners have been identified and listed by Argyle (1994): amount of speech, emotional tone, degree of dominance, intimacy, role relations, and definition of the situation.

Taking action, one of the last steps in the advisory process, can involve many different activities. This may be the purchase and use of a specific product, e.g. a vaccine. It can also imply changes in management, like milk replacer feeding routine. Starting to participate in a regular FHPM or QRM programme together with the veterinarian, or beginning to use computer-based management programmes, also implies taking action.

Although it may seem the easiest part of the process, this step needs careful planning and effective communication once interest and desire were evoked. Opening a 'channel' for action can facilitate the desired and necessary measures to be taken (Bertrand *et al.*, 2007).

What does this mean? The client's behaviour can be interpreted as a dualism. The preference, of the status quo, and the reluctance to change a well-known (sometimes poor functioning) system on the one hand, and the desire to try something new and unusual on the other hand. A simple and uncomplicated plan and a clear schedule from the consulting veterinarian will open a channel for the action and aid the farmer in attempting to adopt a new management.

Inversely, the client may not succeed in adopting the changes, if he is left alone in the early phase. Any problem arising here may block the channel for innovation and further action is not taken. Therefore, a coaching track should be developed in parallel.

Different personalities and situations require different approaches, depending on the relationship between veterinarian and farmer. A general rule is, however, that compliance to a certain option is usually better if it has been developed in a 'participative' discussion rather than being 'imposed' onto the client. Especially risk-taking, entrepreneur-like farmers will rather comply with an approach based on a bilateral activity (Kleen, 2008; Kleen and Rehage, 2008).

FHPM and QRM programmes have to be understood as complex, dynamic and constantly evolving processes, always subject to changes and pressures (Sibley, 2006). It is therefore useful to give discussions in the planning process some shape and direction. The SMART-scheme has proved to be useful in complex situations like these. The acronym stands for the way questions should be asked and plans should be made:

- Specific: The communication should apply to the specific conditions and problems that are dealt with.

- Measurable: Introducing realistic and objective figures helps to focus on the actual problem and prevents disagreements (establish benchmark or farmers' study groups).
- Achievable: Instead of aiming for unrealistic targets, like completely eradicating a problem in a herd, both the consultant and farmer should try to achieve a certain goal that is actually reachable and makes success measurable. In that way, disappointment is avoided and motivation kept high for FHPM or QRM.
- Relevant: The planning should focus on actual, costly and immediate problems rather than spending too much effort on less relevant targets and too low priorities.
- Time-based: In order to make success visible, thereby motivating both veterinarian and farmer, deadlines and fixed evaluations should be used. This will make achievements both visible and objective.

The SMART-scheme can aid in concentrating on the actual problems. It helps preventing waste of time and energy, and in avoiding disappointments.

Most of the communication process is taking place subconsciously and deals with a lot more than just the content of a conversation. A veterinarian should take enough time to analyse the status of the consultation and the role he is expected to play by the client. Acting according to this principle helps in avoiding misunderstandings and prevents unnecessary friction that is disturbing the advisory process.

Acknowledging some basic principles of decision-making and communication is therefore more than a kind of psychological magic. It helps veterinarians improving their standing in competition and helps in creating a more relaxed and more satisfying working atmosphere within the changing cattle industry in general, and in young stock herd health or quality risk management in particular.

Further details about professional communication can be found in Kleen (2008), and in Kleen and Rehage (2008).

Dairy farmers often say that the rearing of young stock is a costly business, most of all because it would regard a non-productive period of about two years. However, we hope to have made clear that young stock rearing is, on the contrary, an highly economic activity. It is about the investment in the future of the dairy farm. Young stock rearing has several important productivity performance areas, like health state, growth rate, reproductive performance and genetic make-up.

Nevertheless, young stock rearing often seems to be a neglected domain of the dairy. Consequences are that animals are inappropriately observed for clinical signs of disorders or for proper daily growth rate. Other problems become manifest when the reproductive period arrives, because puberty may be delayed and body weight insufficient, leading to poor oestrus expression if any. Subsequent pregnancy rates may be too low which results in a late age at first calving.

With this book we hope to have demonstrated that dairy young stock rearing is indeed a business. A well organised business with operational goals and productivity performance parameters to monitor closely, with structured methodologies to follow, and with evaluation procedures which assist in adjusting components in the young stock rearing process.

The better this organisation functions, the structure applied and the evaluation conducted, the less the economic losses due to diseases and disorders will be, and the better the farm income.

In this book, several programmes were introduced: herd health and productivity management, farm health and productivity management, quality risk management. Each of these can be extended with the other components addressed, like cow comfort, and biosecurity plans. The philosophy is to gradually expand a farm health and productivity programme with the forenamed components, and over time conduct an up-scaling to a quality risk management programme with a focus on animal health, animal welfare and public health. In that way, the dairy farm as a whole will be better prepared for the demands from the dairy food production chain, the consumers and society.

Veterinary practitioners have a paramount role to play in this field, but they must be willing to take the challenge, acquire new knowledge and skills, and invest in their own future (Cannas da Silva *et al.*, 2006). Then they will have added value for both the dairy farmer, the dairy food production chain and society.

Annexes to Section I

Annex 1. Common causes of infectious disease in young stock (after Radostits *et al.*, 2000 and 2007; Radostits, 2001a,b; Smith, 2009)

Table A1.1. Diarrhoea in young calves.

Agent	Bacteria	Virus	Parasite	Zoonosis	Age (days)
Septicaemia by *E. coli*	X				< 1 ½
Enterotoxigenic *E. coli* (ETEC)	X				0-7
Rotavirus		X			5-15
Coronavirus		X			5-21
Cryptosporidium parvum			X	X	5-35
Salmonella spp.	X			X (?)	5-42
Clostridium perfringens B and C	X				5-15
Eimeria spp.			X		> 30

Table A1.2. Most common infectious agents causing disease in calves from 2 to 10 months of age.

Agent	Bacteria	Virus	Parasite	Zoonosis
Respiratory				
BRSV		X		
PI-3		X		
BHV-1		X		
BVD/MD		X		
Mycoplasma spp.	X			
Mannheimia haemolytica; Pasteurella multocida	X			
Haemophilus somnus	X			
Salmonella spp.	X			X
Dictyocaulus viviparus			X	
Gastrointestinal				
Ostertagi ostertagi			X	
Trichostrongylus axei			X	
Nematodirus helvetianus			X	
Coopereria oncopohra			X	
Other				
Fasciola hepatica			X	

Table A1.3. Common infectious causes of abortion in cattle.

Agent	Bacteria	Virus	Parasite	Zoonosis
Neospora caninum	X		X	
Brucella abortus	X			X
Listeria monocytogenes	X			X
Chlamydia pecorum		X		X
Leptospira (pomona)/hardjo	X			X
BVD		X		
IBR		X		
Trichomonas fetus	X			
Campylobacter spp. (?)	X			X
Bacillus licheniformis	X			
Arcanobacter pyogenes	X			
Salmonella spp.	X			X

Annex 2. Hazards and risk factors of disorders in each of four age groups

Table A2.1. Rearing period from birth to 2 months of age.

Disorders	Risk factors
Disorders around birth	Bull calf – More muscular breed sire used – Posterior position at birth (milk fever in dam; heavy stress; low vitality) – 1st parity
Diarrhoea in 1st week of age	Wrong or no anti-scours vaccination (resistant bacteria; GVP not applied) – Poor hygiene around birth (too many calving cows in pen; calves born on slatted floor; no attention of farmer; lack of time) – Poor housing hygiene (no attention of farmer) – Poor colostrum quality (unhygienic collection; low IgG level; unhygienic feeding; poor storage practices; using colostrum frozen more than 1 year ago; dilution with water; poor thawing practices) – Too long interval between birth and 1st meal – First colostrum milked is not given to calf – Insufficient quantity is given to calf each meal – Too early group housing (too little space; no individual pen) – Inside housing – No preventive antibiotics – No free choice salt is provided to calves – Large herds – Too little attention of care-taker – Sudden changes in feeding practice – No routine monitoring visits by veterinarian – Heat stress in calves
Diarrhoea in older calves	Use of maternity pen as sick pen – No individual calf hutches – Poor rearing hygiene practice (improper housing and feeding; damp bedding material) – Milk replacer fed without antimicrobials – Sick calves not isolated – Drinking water pH >8 – Roughage from plots where manure was spread without ploughing under – Sudden changes in feeding or ration – No attention of care-taker
Navel disorders	Navel haemorrhage (internal or external) due to inadequate stretching of umbilicus e.g. after C-section – Infected navel (dirty calving pen; dirty calf hutch; no navel disinfection applied; navel suckling by others; calf born on slatted floor) Navel hernia (genetic cause? sex effect? twin birth) – Navel cord too short (poor manipulation during birth eg C-section; posterior position at birth) – Visceral organ eventration through navel (genetic background?)
Poor weight gain	Infection in calf (navel disorder; diarrhoea; respiratory) – Feeding poor quality colostrum (see under diarrhoea 1st week) or hay – Feeding poor quality milk replacer (wrong composition; poor quality water used at mixing; milk powder not stored dry; wrong preparation or supply (temperature; poor mixing; dirty conditions; wrong feeding temperature) – Stressful calf-handling – No or poor water quality – Too much hay or grass given – *Ad libitum* milk replacer given (hampers intake of hay = poor rumen development=stress and susceptibility after weaning) – Wrong concentrates composition – Feeding mastitic or antibiotic milk – Stress/infection at dehorning (poor dehorning practice; wrong timing of dehorning; stressful handling; too small headspace in feedrack) – Lack of concern of care-taker

Table A2.1. Continued.

Disorders	Risk factors
Respiratory disorders	Poor barn climate (temperature, humidity, draughts; not adapted to meteo changes) – Housing older calves with younger, or too many age groups mixed – Ammonia air level too high – Lack of attention of care-taker – Season – Poor colostrum management earlier – Poor record keeping practices – Animal density too high – BVD infection present in herd – Poor bedding material quality – New cattle are purchased – Grazing in summer – Previous disease (diarrhoea; respiratory)
M. paratuberculosis infections	Unhygienic conditions around birth (calving pen not cleaned and disinfected) – Group maternity pens – Unhygienic birth – Calf not removed from dam immediately (no supervision; bull calf; other work) – Calf born on slatted floor – Poor colostrum management (see above) – Borrowed colostrum –Transplacental infection from infected cow – Purchased infected dam – Waste milk feeding – Contacts with older cattle
Wrong identification of calf	Several cows calve at the same time – Calves born at the same time in the barn or in pasture – Mixing up ear tags – Poor record keeping – Ear tag of dead calf placed in another calf – Twins with unknown dam – Computer crash and loss of data
Wrong extra teat removed	Poor record keeping – Wrong identification of teat to be removed – Wrong teat removal practice

(after Radostits, 2001a,b; Boersema, 2006; Smith, 2009)

Table A2.2. Rearing period from 2 to 15 months of age.

Disorder	Risk factors
Poor weight gain	Infection causing disease in calf (navel infection; ear infection after tagging; diarrhoea; respiratory disease) – Overcrowding – Quality of feed or water too low – Providing insufficient energy/minerals – Too low light intensity in barn – Wrong weight at insemination
Respiratory disorders	Herd-density too high – Care-taker too busy – Poor barn climatic conditions – Sudden weather change – Poor animal management (no hair clipping at the beginning of barn period) – Poor grazing management – Poor lungworm infection prevention strategy
Diarrhoea in older calves or maiden heifers	Herd prevalence (e.g. salmonellosis) – Farm size – *Fasciola hepatica* infection – Feeding roughage from fields where manure was applied and not ploughed under during the same growing season – Poor hygiene practices in calf rearing (damp bedding) – No isolation of sick animals – Offering water with pH >8.0 – Lack of concern of care-taker – Feed related – Sudden change of diet
Locomotion problems	Poor claw trimming – Overcrowding – Poor hygiene in cubicles – Too little space in cubicles – Older animals – Inside housing/zero-grazing – Large herds – Previous Mortellaro infections – Animals in heat
Gastro-intest. worms	Poor grazing management – Poor de-worming strategies
Liverfluke infections	Animals grazing on wet land (presence of ditches; high ground water level) – Presence of snail habitats – Poor de-worming
Trichophyton verrucosum	Contact with other infected animals – Poor hygienic practices – Wrong treatment of infected animals – Poor climatic conditions (too high humidity; high temperature) – Purchase of infected cattle
Fattening of udder tissue (± 8 mo age)	Feeding too high energy rations between 3 and 10 months of age
M. paratuberculosis inf.	Contacts with adult cattle or with manure before 1 year of age – Insufficient separation of calves from adult cattle
Not pregnant < 15 months of age	Lack of proper heat detection – Wrong bull selection – Wrong insemination technique – Lack of mount detection – Heifer not in oestrus before 15 mo of age – Light intensity too low (less oestrus behaviour signs) – Anoestrus/suboestrus – Wrong body weight for insemination

Table A2.3. Rearing period from 15 months of age to 3 weeks prior to the 1st calving.

Disorder	Generic checklist of potential risk factors as derived from literature
Abortion	Herd prevalence of several abortion causing agents – Large farms – Buying heifers from other farms – Grazing heifers on pasture next to infectious animals from neighbouring farm(er) – Lack of vaccination pre-breeding (e.g. BVD, IBR) – Not in a certified herd health scheme (e.g. BVD, salmonellosis)
Poor weight gain	Wrong weight at insemination – Infection causing disease in calf (navel infection, ear infection after tagging, diarrhoea, respiratory disease, gastrointestinal worms) – Overcrowding – Quality of feed or water is too low – Providing insufficient energy and minerals
Lameness	Poor claw trimming practices in young stock – No routine screening of claws in heifers – Overcrowding – Bad hygiene in cubicles – Too little space in cubicles – Older animals – Inside housing/zero-grazing – Large herds – Feeding management failures
Respiratory disorders	Too high herd-density – Care-taker too busy – Poor barn climatic conditions – Sudden weather change – Housing does not allow proper ventilation – Poor grazing management – Wrong lungworm infection prevention strategy
Gastro-intest. worms	Poor grazing management – Wrong de-worming policy – Grazing heifers which have never been outside in the preceding season
Trichophyton verrucosum	Contact with other infected animals – Poor hygiene practices – Wrong treatment of infected animals – Poor climatic conditions in the barn (high humidity; high temperature)
Liverfluke	Animals grazing on wet land (presence of ditches; high ground water level) – Presence of snail habitats – Wrong 'de-worming' policy
Mastitis	Teat/udder-sucking – Flies in cubicles – Milk leaking before calving – Poor general hygiene – Poor animal hygiene
Overconditioning	Too high energy rations fed between 3 and 10 months of age
Heifer comfort	Too small cubicles compared to body size – Introduction of heifers in herd (rank fights) – Poor hygiene in calving pen and cubicles – Health threats like mastitis, lameness and abortion – Poor adaptation to 'lactation-rations' – Poor climatic conditions (high humidity; high temperature) – Feeding management failures – Poor quality of feed and or water

Table A2.4. Rearing period: around 1st calving.

Disorder	Generic checklist of potential risk factors as derived from literature
Dystocia in heifer	Bull with low calving ease index – Heifer body condition < 3 or > 3 ½ (wrong composition (close up) rations or caused by diseases – 1st parity heifer – Younger age at calving (<22 month) – Calf relatively too large – Poor social interactions with herd mates – Milk fever/loss of feed intake
Retained placenta	Over-conditioned at calving – Stress around calving – Assistance during labour – Decreased general immune-responsiveness (BVD, ketosis)
Trauma in newborn calf	Dystocia of dam – Too small calving pen – Stressed dam (steps on calf) – Born on slatted floor – Born in pasture without supervision too close to ditch – Poor management intervention – Imprudent assistance with calving jack
Wrong identification of newborn calf	More cows/heifers calving at the same time – Calves born without supervision in close up group or in pasture – Wrong recording – Mixing up ear tags – Tag of dead calf placed in other living calf so calf can be removed sooner from farm – Computer crash without backup
Mastitis	Teat/udder-sucking – Flies – Poor hygiene in cubicles – Milk leaking *prepartum* – Milk fever – Blood in milk – Teat oedema – High peak milk flow – Teat canal protrusion – Over-conditioning of heifer
Milk fever	Higher parity – High milk production level – High body condition score – Positive cation-anion balance in diet – High calcium level in feed – Season – Climate – Housing – Lush pastures – Lack of exercise – Length of dry period – *Prepartum* milking –Loss of feed intake
Displaced abomasum	Negative energy balance *prepartum* and subsequent ketosis – High body condition score – Suboptimal feed bunk management – *Prepartum* diets containing >1.65 Mcal of net energy for lactation/kg of dry matter – Winter and summer season – Rapidly changing weather conditions – High genetic merit – Low parity – Milk fever – Puerperal metritis– Mastitis – Stress – Non social adaptation of heifers in milking herd
Lameness	Poor claw trimming – Overcrowding – Large herds – Poor hygiene in cubicles – Too little space in cubicles – Inside housing (on slatted floor)/ zero-grazing – Unequal or loose slats – Floors (slatted or not) without dung remover – Feeding failures (ration formulation, ration changes)
Udder oedema	Positive cation-anion balance in ration during late gestation – 1st parity heifer – Too high protein level in ration
Respiratory disease	High herd-density – Care-taker too busy – Poor ventilation and barn climate – Sudden weather change – Housing does not allow proper ventilation – Poor grazing management – Wrong lungworm infection prevention strategy

Table A2.4. Continued.

Disorder	Generic checklist of potential risk factors as derived from literature
Liverfluke	Animals grazing on wet land (presence of ditches; high ground water level) – Presence of snail habitats – Wrong 'de-worming' policy
Poor heifer comfort	Too small cubicles compared to body size – Introduction of heifers into milking herd (rank fights) – Poor hygiene in calving pen and cubicles – Health threats like mastitis, lameness and abortion – Poor adaptation to lactation-rations – Poor climatic conditions (high humidity; high temperature) – Feeding management failures – Poor feed and or water quality – Excessive social interactions – Proper transition into lactation not achieved
Poor feed intake	Poor quality of feed and water offered – Overcrowding/rank fights – Discomfort due to disease or heat stress (Temperature humidity Index) – Dystocia and *postpartum* problems – Negative energy balance/ketosis – Over-conditioning – Leaving calf with dam.

Table A2.5. Summary of the types of hazard for each of the four different age-groups.

Period	Major hazards			Management		
	Animal health (AH)	Animal welfare (AW)	Public health (PH)	Feeding (FM)	Recording (R)	Other
I (0-2 mo)	• Diarrhea in neonate • Navel disorder • Respiratory disorder	• Birth problems • Dehorning	• *Salmonella* • *Cryptosporidium* • Para TBC [1] • *E. coli* O_{157}	• Poor weight gain	• Wrong identification calf • Lacking growth recording calf	• Wrong extra teat removed
II (2-15 mo)	• Respiratory disorder • Gastrointestinal disorder • Lameness	• Poor locomotion	• *Trichophyton verrucosum* • *E. coli* O_{157} • Para TBC [1]	• Fattening of udder tissue • Poor weight gain	• Lacking mount/AI recording	• Heifer not pregnant before 15 mo of age
III (15 mo-3 weeks *prepartum*)	• Death of foetus • Lameness • Mastitis	• Heifer comfort loss	• *Trichophyton verrucosum* • *E. coli* O_{157}	• Poor weight gain	• Lacking growth recording heifer	• Abortion and (economic) loss of heifer as a result
IV (-calving)	• Mastitis • Lameness	• Heifer comfort loss	• *E. coli* O_{157}	• Poor feed intake (incl alimentary disorders)	• Wrong identification of newborn calf	• Trauma calf

[1] *Mycobacterium paratuberculosis* in cattle has certain similarities to Crohne's Disease in humans and therefore is sometimes thought to be potentially dangerous for public health, but the causal relation has still not been established. Paratuberculosis cannot affect public health in the period calf-milk heifer, because clinical disease does generally not occur until 3 or 5 years of age. Although infection in calves takes place in the 1st year of life, therefore precautions in calf rearing management can prevent animals being infected with paratuberculosis (Radostits *et al.*, 2000 and 2007). Hence, in severely affected herds, paratuberculosis may also spread horizontally among calves; this may cause the calf care-takers to occur the risk of being infected or at least having contact with the bacteria (GD personal communication, 2009).

Annex 3. Hazard (disease) occurrence over time in each of four different age groups

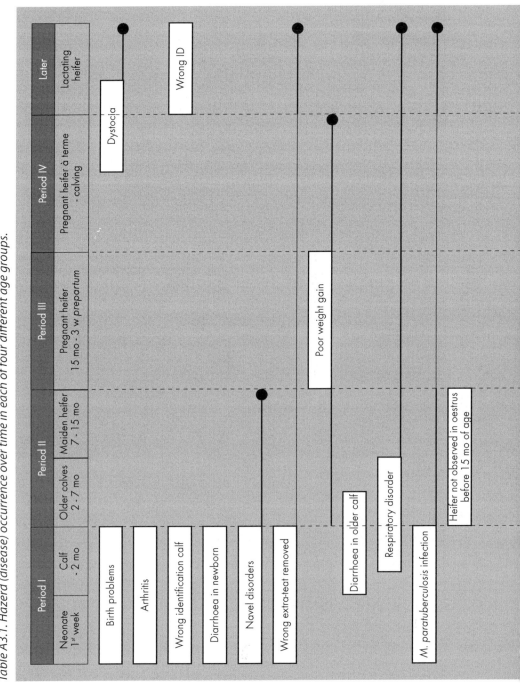

Table A3.1. Hazerd (disease) occurrence over time in each of four different age groups.

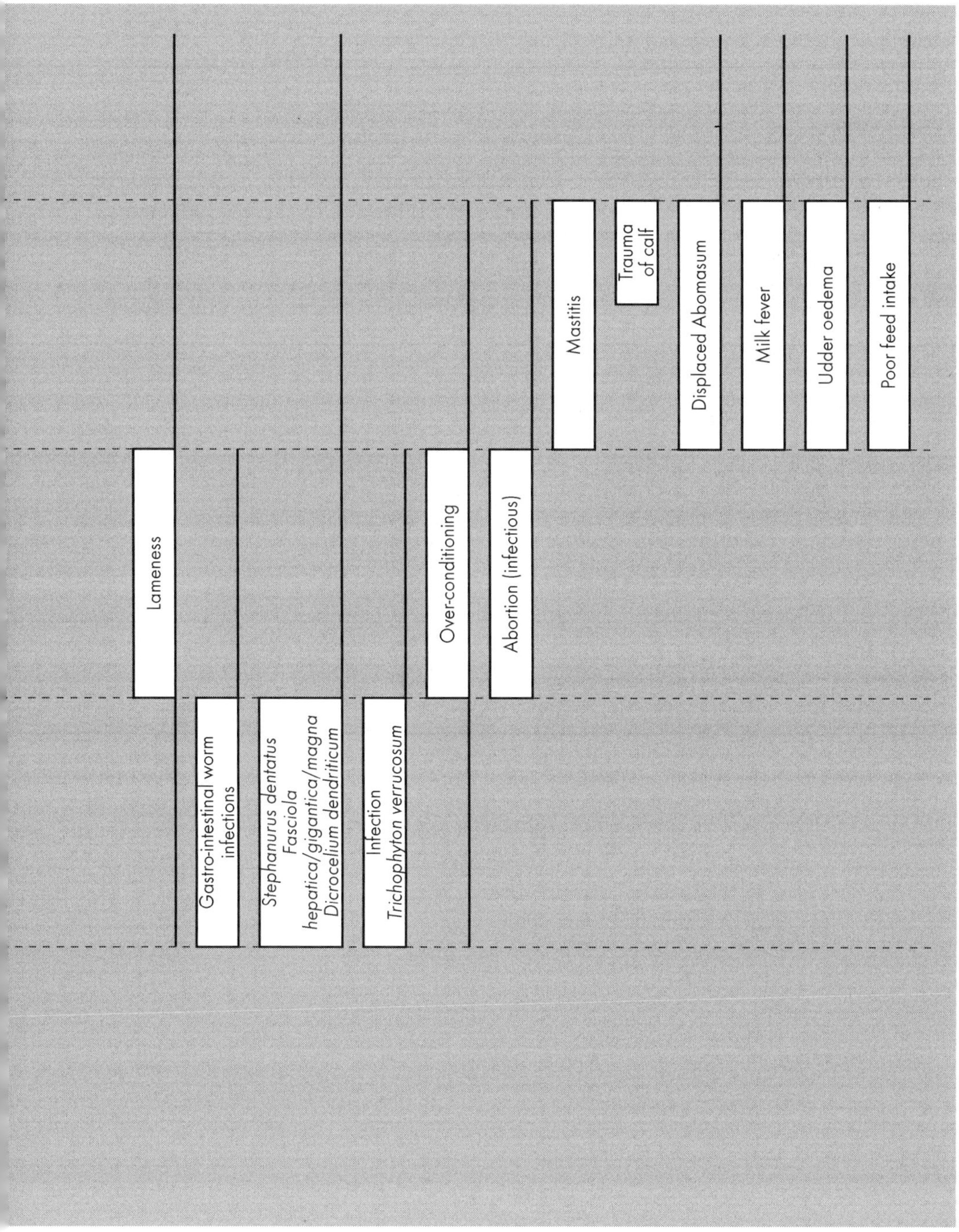
Lameness
Gastro-intestinal worm infections
Stephanurus dentatus
Fasciola hepatica/gigantica/magna
Dicrocelium dendriticum
Infection
Trichophyton verrucosum
Over-conditioning
Abortion (infectious)
Mastitis
Trauma of calf
Displaced Abomasum
Milk fever
Udder oedema
Poor feed intake

Annex 4. Young stock treatment advisory plan: an example

Table A4.1. Example of young stock treatment advisory plan.

Clinical signs	Probable disease	Further diagnostics	Degree of illness
Watery diarrhoea, 'sunken eyes', T: >39.5/<38 °C listlessness	*E. coli* diarrhoea in newborn (0-7d)	Laboratory - faeces - necropsy	Whenever sick Severely weak calf
Yellow paste-like or more liquid diarrhoea, weak and listlessness	Rota-Corona diarrhoea in newborn (7-14d)	Laboratory -faeces	Severely weak calf
Yellowish, paste-like faeces	Feed-related diarrhoea		Diverse
Bloody diarrhoea with pieces (tissue), listlessness, respiratory disorder and abortion in cows T: >39.5 °C!	*Salmonella* diarrhoea in older calf (>3weeks)	Laboratory - faeces - necropsy	More animals affected 1 (severe) sick calf
White-yellow-green, watery, maybe blood	Cryptosporidiosis	Laboratory - faeces	
Brown-green, thin, often blood. Calf straining and poor weight gain.	Coccidiosis	Laboratory - faeces	

Treatment					Follow up *Short term* **Long term**
Drug	Application	Dose/freque	Withdrawal time	Complementary	
Contact your veterinarian immediately					*Recovery within*
Quinolones (Cephalosporines if sepsis)	IV, IM IM, SC	See label	See label	Provide with as much electrolytes as possible	*few hours-within half a day* **Follow work instruction 1**
-	-	-	-	Electrolyte therapy 1 day	*Recovery within one day*
Quinolones Cephalosporines if sepsis	IV, IM IM, SC	See label	See label	Provide with as much electrolytes as possible	**Follow work instruction 2**
No treatment!!				Feed 2 times half amount of milk. Thereafter increase amount gradually.	**Follow work instruction 3**
Contact your veterinarian immediately					**Follow**
Quinolones Cephalosporines if sepsis	IV, IM IM,SC	See label	See label	Provide with as much electrolytes as possible	**biosecurity plan -** *Salmonella*
Contact your veterinarian immediately					
(Halofuginone) (Disinfection with sodiumhypochlorite, potassiumperoxymonosulphate; or formalin fumigation if permitted)					
Sulfadimidine-Na Diclazuril	Oral Oral	See label	See label	Electrolytes in case of dehydration + NSAID's in necessary	**Follow work instruction 1**

Table A4.1. Continued.

Clinical signs	Probable disease	Further diagnostics	Degree of illness
Brown-grey clay-like or water thin brown. Calf has splashing sounds in abdomen	'Ruminal drinker'	-	
Cough, respiration rate↑, dirty nose	Acute respiratory disorder (viral or/and bacterial)	Laboratory - necropsy	
			Severe weak calf
			10-15% calves sick
Swollen, warm and painful umbilicus cord	Infection umbilicus	-	
Swollen, warm and painful joints, lameness	Arthritis	-	
Deep/harsh cough (after exercise), respiration rate↑, anorexia and poor weight gain *Only calves grazed outside or fed infected fresh grass are at risk!*	Respiratory disorder (lungworm)	Laboratory - serology: IgG - faeces: larvae or eggs	All animals preventive
			Animals with clinical signs
			Severely sick calves
Lameness	Disease of Mortellaro	-	
	Laminitis		
	Interdigital dermatitis		

Treatment					Follow up *Short term* **Long term**
Drug	Application	Dose/freque	Withdrawal time	Complementary	
Early: empty rumen with stomach tube and provide teat bucket feeding Chronically: wean calf					**Follow work instruction 2**
Procaine benzylpenicilline Florfenicol Cephalosporines	IM SC,IM IM,IV	See label	See label	Provide Bronchodilatation + NSAID's + glucocorticosteroi	**Follow work instruction 3**
Quinolones Florfenicol	IV, IM IM,SC	See label	See label	Provide with as much electrolytes as possible	

Contact your veterinarian immediately

Drug	Application	Dose/freque	Withdrawal time	Complementary	Follow up
Procaine benzylpenicilline	IM	See label	See label	NSAID's	**Follow work instruction 2**
Trimethoprim/S Ampi-/ amoxicilline Benzylpenicilline/ neomycine	IM	*See label*	*See label*	Veterinarian could flush joint with ringer lactate solution + NSAID's	**Follow work instruction 3**
Specific anti-lungworm drug	Topical	See label Once in autumn and once 2 weeks after housing	See label		**Follow biosecurity plan - parasitic infections + follow work instruction for grazing calves**
Ivermectine	SC	See label	See label		
Ivermectine	SC	See label	See label	NSAID's AB	
OCT-spray Cephalosporines Ration adaptation Risk management Claw trimming Risk management	Local IM, SC	See label	See label	Claw care!! NSAID's	**Follow biosecurity plan - lameness**

Table A4.1. Continued.

Clinical signs	Probable disease	Further diagnostics	Degree of illness
Dry, rough and dull hair, poor weight gain, anorexia, listlessness and (intermittent) diarrhoea. **Only calves grazed outside or fed infected fresh grass are at risk!**	Gastrointestinal worm infection	Laboratory -blood	All animals preventive
			Animals with clinical signs
			Severe sick calves
Chron.: poor weight gain Acute: submandibular oedema and pale mucous membranes	*Fasciola hepatica*	Laboratory - blood - necropsy	
Restless, perineum wet and bloody, placenta appears in vulva,	Abortion	Laboratory - blood - necropsy	
Round grey-white crusts	Infection *Trichophyton verrucosum*		

Note: Biosecurity plans and work instructions, as mentioned in this Annex, refer to specifically designed action plans. They are not further elaborated; see the respective chapters.

Treatment					Follow up *Short term* **Long term**
Drug	Application	Dose/freque	Withdrawal time	Complementary	
Specific anti-lungworm drug	Topical	See label Once in autumn and once 2 weeks after housing	See label		**Follow biosecurity plan - parasitic infections + follow keynote grazing calves**
Ivermectine	SC	See label	See label		
Ivermectine	SC	See label	See label	NSAID's AB	
10% triclabendazol	Oral	See label Once in autumn and once 2 weeks after housing	See label	Symptomatic f.e. NSAID's if necessary	**Follow biosecurity plan - parasitic infections**
Contact your veterinarian immediately					**Follow biosecurity plan - abortions**
Enilconazol Econazol (vaccination)	Topi-cal	See label	See label	Remove harsh crusts first	**Follow biosecurity plan - *Trichophyton verrucosum***

Annex 5. An example of a risk management procedure in young stock rearing: diarrhoea (after Boersema, 2006)

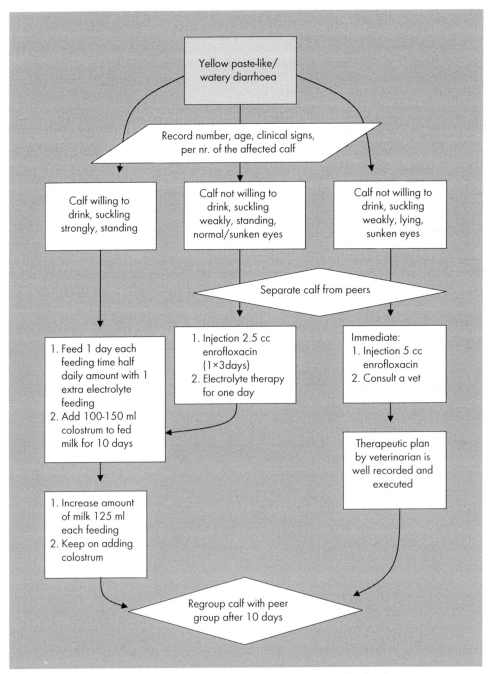

Figure A5.1. Example of working instruction 'Diarrhoea in neonatal calves'.

Annex 6. Good hygiene in neonatal calves: worksheet on hygiene in the calving pen and around calving

Company:	Farm X	Date last revision:	21/06/2009
Editor:	xx xxxx	References:	Dutch Dairy Manual (2006)
Resp. person(s):	xxxxxx xx		Van Trierum (2005)
	xxxx xxx xxxxxx		Brand *et al.* (1996)
Aim:	Optimal hygiene of neonatal calves		

Contents

- Hygiene: cleaning calving pen: what-where-when-who-why
- Hygiene during delivery: person, materials used
- Animal care before/after calving: remove udder hair, time to standing heifer, 'breast-belly' position calf, disinfect naval, remove calf, feeding/drinking heifer, regrouping with herd, etc.

Hygiene

- Clip dams (cows, heifers) before calving and remove udder hair too.
- Clean pen directly after fresh cow re-entered the herd following protocol 'golden standards for cleaning and disinfection'.
- Use enough fresh straw for bedding.
- As long as the same cow is in the pen, add new fresh straw when bedding gets dirty.
- Make sure the bedding is clean when cow/heifer starts calving.
- Clip cows/heifers before calving and remove udder hair.
- Never house sick cows together with pregnant cows/heifers.

Hygiene during delivery

- Store materials, equipment and disinfecting solutions in a fixed place. Make a checklist containing all things necessary for adequate assistance.
- Person who helps during delivery must wash his/her hands properly with substance from list B in Annex 8.
- Assistents should wear clean boots and overalls or specially designed suits.
- Materials used during the assistance must be clean at the start and kept in bucket with dilution of substance from list B in Annex 8.
- Perineum and tail must be washed with the same dilution B in Annex 8.

- Make sure hands are always clean, or better still: wear gloves, before exploring the vagina of the cow.
- Clean, disinfect and dry materials and equipment after delivery and store them in the place they belong. Make sure everything is in place (and everything is present in sufficient amounts).

Animal care before/after calving

- Cow or heifer should stand within 5 min. after delivery.
- Calf must be removed immediately after birth.
- Umbilical cord should be tied off if excessive haemorrhage is present, and disinfected by dipping the cord in e.g. chlorhexidine or iodine solution.
- Within one hour after birth cow/heifer should be milked for the first time.
- Heifer/cow is not regrouped until dry matter intake is sufficient and no other clinical illness-signs occur. Daily rumen-fill score, faeces consistency and fibrosity scores, temperature measurement, and appearance are conducted and results recorded.
- House cows/heifers in a way that social (visual) contact with herd mates is possible.

Annex 7. Good hygiene in older calves/heifers: worksheet on hygiene in the calf/heifer house

Company:	Farm X	Date last revision:	21/06/2009
Editor:	xx xxxx	References:	Brand *et al.* (1996)
Resp. person(s):	xxxxxx xx		Dutch Law
	xxxx xxx xxxxxx		Dutch Extension Service
Aim:	Prevent poor hygiene-related diseases		

Contents

- Calves: General hygiene practice
  Single housing: hygiene (cleaning, 'w-w-w-w-w', etc.) and housing
  Group housing: hygiene (cleaning, 'w-w-w-w-w', etc.) and housing

(w-w-w-w-w = what-when-where-who-why; see Annex 6)

General Hygiene Practice (GHP)

- Calf caretaker should wear his boots and overall only in calves department.
- Work with clean hands; even better is wearing gloves.
- Equipment must be clean before use ((teat)buckets, thermometer, measuring cup and stirrer).
- Oesophageal feeder should be cleaned and disinfected between calves.
- After feeding equipment must be cleaned and dried.
- Feed calves, moving from young to old.

Housing

- Calves must be housed single during the first 10 days of life as soon as possible after birth.
  - Before calves are moved to their single box, make sure this pen was cleaned and disinfected (following 'Worksheet Golden Standards for Cleaning and Disinfection', Annex 8).
  - Single pens should measure as wide as wither height and as long as 1.1 times body length (according to Dutch-law).
- After 10 days calves should be housed in larger pens where social interaction with calves of the same age is facilitated.
  - Never house calves differing more than two weeks in age, in the same pen.
  - Never allow contact between pre-weaned and weaned calves.

- Do not house calves on slatted floors and cubicles before 10 weeks of age.
- Before moving a new group of calves in a group-pen, make sure this pen is cleaned and disinfected (following 'Golden Standard').
- Calves in group-pens need 2-4 m² space (1.5-1.8 m² in Dutch Animal Welfare Law).
- Check 2 times daily if the pen bedding is still clean and dry. If not, adapt the bedding material.

Annex 8. Good cleaning and disinfection: golden standards for cleaning and disinfection

Company:	Farm X	Date last revision:	21/06/2009
Editor:	xx xxxx	References:	Brand *et al.* (1996)
Resp. person(s):	xxxxxx xx		
	xxxx xxx xxxxxx		
Aim:	Prevent poor hygiene-related diseases		

Contents

- Golden standards for cleaning and disinfection
- List A: disinfection solutions
- List B: antiseptic solutions

Golden standard for cleaning and disinfection:

1. Remove straw and manure.
2. Clean walls and floor with high pressure cleaner.
3. Disinfect walls and floor with one of substances from list A.
4. Rinse and let it dry.

List A: disinfection solutions containing one or more of these substances:

- Alkyldimethylbenzylammoniumcloride
- Didecyldimethylammoniumchloride
- Glutaraldehyde
- Isopropanol
- Formaldehyde
- Isopropanol
- Natrium-p-tolueensulfonchloramide

List B: antiseptic solutions used for disinfection of skin:

- Iodine
- Chlorhexidine
- Alcohol

Note: for prevailing infections with *Cryptosporidium* spp. among young stock, it is advised to thoroughly clean and disinfect the premises with sodiumhypochlorite or potassiumperoxymonosulphate. Formalin fumigation can be effective but is carcinogenic, so at least gloves and mask must be put on when applying, if application is allowed by law at all.

Annex 9. Major management elements in neonatal calf care

There are two paramount *performance parameters for calves in their early life*: (1) health status, (2) daily growth rate and development. In the subsequent life stages, several other factors may influence this calf performance.

A9.1. Effect of the dam

The *dam* may influence the health status of the new-born calf in several ways; the main phenomena are given here (Correa *et al.*, 1993; Baumgartner, 1999; Chassange *et al.*, 1999; Staufenbiel *et al.*, 2003; Noordhuizen, 2004):
- When they have a (too) high body condition score at the end of the dry period.
- When, in advanced pregnancy, they have fatty liver disease.
- When no prophylactic measures were applied in the dry period, like vaccination during the last two months of pregnancy against *E. coli*, Rotavirus and Coronavirus, to give to the calf, through the colostrum, a good protection, called passive immunity.
- When they suffer from hypomagnesaemia or hypocalcaemia.
- When they experience a ketosis at late pregnancy (due to e.g. a disease like severe mastitis), causing lower antibody levels transmitted through the colostrum (i.e. poor colostrum quality). Colostrum quality can be measured semi-quantitatively by a colostrometer (Figure A9.1).
- When they show low Se and vit E levels around calving (see Section II for reference values).
- When they are affected by subacute rumen acidosis (SARA).

Figure A9.1. Colostrometer in a tube with milk. The reading is as follows: red= insufficient (<20 mg/ml); yellow= doubtful (20-50 mg/ml); green= sufficient (> 50 mg/ml). The colostrometer is a lactodensimeter; read-outs at 22 °C or 72 °F.

- When their dry period Cation – Anion – Balance is too high (causing udder oedema in heifers and dilution of colostrum in second and higher parities leading to less quality).

Some addresses to order a colostrometer:
- Biogenetics; 09359 Highway 126. Mapleton, Oregon 97453 USA.
- CalfGard. North Field Laboratories Pty Ltd.; 180 Fosters Rd, Oakden, South Australia 5086, Australia.
- Colostrum Densimeter, www.kruuze.com.

A9.2. Birth management

There are several 'golden rules' to apply when *managing the birth of the calf* and the health of the dam. Some of these rules are given here:
- The place of calving must be clean, fresh, not slippery, comfortable, quiet and separated from the other cows of the herd, but herd mates must be visible and heard. It is strongly advised to have a special maternity pen.
- The perineum of the dam must be thoroughly washed when calving starts.
- Contamination of the calf by the dam or others must be avoided, and a proper biosecurity plan at calving/birth should be applied to achieve an acceptable level of safety.
- If the cow or heifer needs assistance during calving, all the correct managerial and hygiene procedures must be applied: Using gloves (unless proper disinfection with chlorhexidine has taken place) and being patient during all procedures, are just two major issues when we are dealing with a *dystocia*.
- Decisions must not be delayed, for example a caesarean section must be performed at the right time, or a uterine torsion must be solved in time and according to the state of the art.
- A long-lasting intervention must be avoided because it may induce calf acidosis.
- Lay the calf in a lateral position after birth; dry it and apply thorough massage.
- The navel must be externally disinfected with an antiseptic spray immediately after birth, in a proper way.
- Clear the calf of its meconium when it is obstructing the rectum; also eliminate obstructions of the oropharynx (e.g. by hanging it upside down).
- Separate the calf from its dam and bring it into a separate unit with single pens
- Give the calf freshly and hygienically milked colostrum several times a day (see details later on). Disinfect the material used for colostrum supply.

For several of these golden rules it could be advisable to design technical, *management work instructions*, which usually do not comprise more than 1 page DIN A4. Such work instructions are supportive for both the farmer, farm workers and the veterinarian.

For example, a vaccination scheme is a work instruction; what to do in case of dystocia is another one.

A9.3. Immediately after birth

Immediately after birth, there are different activities to undertake. These activities should be embedded in a strict routine on the farm. Again, they may be integrated in a work instruction called 'Routine activities after birth'. These activities comprise:
- Separate the calf from its dam, and bring it to a separate unit with single pens.
- Cleaning the calf of fluids from its dam, with movements from tail to nose.
- Not touching the calf with the hands inside its mouth.
- Disinfecting the whole outside of the navel with an appropriate solution (e.g. iodine or chlorhexidine).
- Following strictly the colostrum work instruction on quality control, timings and quantities of meals, on the first and following days. Colostrum quality can be measured by a colostrometer (see Figure A9.1 in this Annex).
- Housing of the calf in a clean, fresh, dry and separate box in a separate barn.
- Calves with diarrhoea must be removed from the calf house and treated separately.
- Biosecurity measures must be implemented (see Section I of this book).

Annex 10. Colostrum management issues

A10.1. Adequate colostrum and optimum newborn calf management

> Begins with the provision of optimal nutrition to the pregnant dam, which will result in a vigorous newborn animal and adequate quantities of colostrum. Ensure optimal health care for the dam to prevent mastitis in the dry period which reduces colostrum IgG content.
>
> At the time of parturition, surveillance of the calving cows and the provision of any obstetrical assistance will ensure that the newborn are born with as much vigor as possible.
>
> The next most important control measure is to ensure that liberal quantities of colostrum are available and ingested within minutes and no later than one hour after birth. While the optimum amount of colostrum which should be ingested at a certain time after birth is well known (Sellers, 2001), the major difficulty with all species under practical conditions is to know how much colostrum a particular neonate has ingested. Because modern livestock production has become so intensive, it is imperative that the animal attendants make every effort to ensure that sufficient colostrum is ingested by that particular species.

The colostrum quantities advised by different authors are unfortunately not always the same. For example Andrews *et al.* (2008) and Godden *et al.* (2009) advise the amount of colostrum to be at least 2 litres in the first 6 hours of life (within 1 or 2 hrs is however more preferable) and two more additional litres in the first 12 hours. The first colostrum should be administered as quickly as possible after birth (within minutes or within 1 hour after birth) to allow a good absorption of the gammaglobulins (IgG) and to give the calf sufficient passive immunity to protect it from viral and bacterial infections, that can affect the intestine and induce diarrhoea. For the first feed, farmers should offer calves up to 3 l of colostrum by nipple-bottle or nipple-bucket and where calves fail to voluntarily consume this volume the remaining colostrum should be fed by oesophageal tube feeder (Godden *et al.*, 2009). See also reference values for colostrum quality below. See Figure A10.1a,b for examples of oesohagal tube feeding.

- When pooled colostrum is used, calves should be fed 3 to 4 l of first-milking colostrum.
- The stomach pH will be around 5 which reduces the incidence of *E. coli* diarrhoea.
- Calves should be fed regularly and preferably by the same person. The dairy calf should be removed from the dam immediately after birth, and put in a single box; they stay there until weaning.
- Colostrum should be milked from the cow aseptically, and 2½ l fed by nipple bottle, teat-bucket or by stomach tube as soon as possible after birth (at least

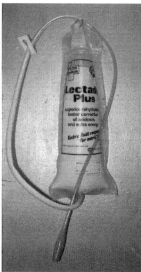

Figure A10.1a,b. Examples of an oesophagal tube for forced colostrum feeding of calves that do not want or are to weak to drink by themselves.

within 1 hour). A colostrometer can be used to check the colostrum-qualtiy and to determine whether a calf is fed a sufficient amount of IgG. Whether a high IgG level has been reached in the calves serum, can be checked through refractometer readings on blood samples (see Figure A10.2a,b).

- Adequate housing and ventilation must be provided to avoid climatic and social stress.

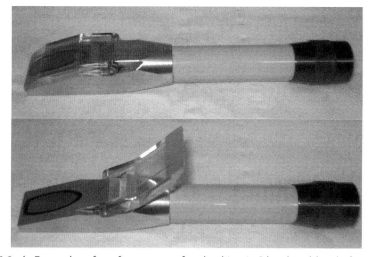

Figure A10.2a,b. Examples of a refractometer for checking IgG levels in blood of young calves.

Some commercial milk – or colostrum – derived oral supplements containing immunoglobulin are available for newborn calves in which colostral intake is suspected or known to be inadequate (Table A10.1). However, they contain low immunoglobulin concentrations compared to those found in high quality first-milking colostrum (Mee and Mehra, 1995). Calves fed whey protein concentrate as a colostrum substitute and administered as a single feeding, was ineffective in preventing neonatal morbidity and mortality compared with a single feeding of pooled colostrum (Mee *et al.*, 1996). See also Annex 11 in this book on specific features of milk replacer products.

A final remark refers to *colostrum replacement products* and *colostrum supplements*. Although their effect may be (slightly because large quantities are needed) beneficial in situations where dam's colostrum or deep-frozen colostrum stock is not available, or of too poor quality, there is a potential risk associated with these products that they might transfer certain pathogens such as *M. avium paratuberculosis, Coxiella burnetii* or *BVD* virus (Mestdagh *et al.*, 2008). On the other hand, certified paratuberculosis-free products may be used in paratuberculosis-affected herds to replace maternal high risk colostrum in order to reduce the spread of disease to the neonate.

In order to eliminate such pathogens – also from colostrum – the heating of colostrum has been proposed (Philippe *et al.*, 2008). Pasteurisation inactivates the pathogens but – on the other hand- reduces the quantity of IgG in the colostrum by about 30%. Heat treatment of colostrum in small quantities at 60 °C during 2 hrs can overcome that problem, while heat treatment at 60 °C during 30 min appeared to be sufficient to eliminate *M. bovis, L. monocytogenes, E. coli* and *S. enteritidis*, but for *M. avium* paratuberculosis a much longer heat treatment would be necessary. In general, a heat treatment of 60 min at 60 °C is recommended to strongly reduce the bacterial load of colostrum (Philippe *et al.*, 2008).

Table A10.1. Reference values for colostrum quality, in g per l serum or in mg/ml milk respectively.

Serum IgG	Interpretation	% of total protein	Milk IgG	Interpretation
5	very poor	okay if		
5-10	insufficient	at 60±7	<20	insufficient
10-15	sufficient	at 48 hrs	20-50	doubtful
>15	good	after birth	>50	sufficient

Annex 11. Assessing calf milk replacers and whole cows' milk (adapted after Rydell, 2002, 2008 and Kunz, 2008)

In choosing milk replacer for the calves, one should pay attention to particular issues. First, to their protein content, energy content, vitamins supplementation, and, possibly, medication. Secondly, to the protein source and their suitability for calves at a certain age.

'The single best criterion for evaluating a calf milk replacer is calf performance.'

(Rydell, 2008)

The *energy and protein* needs in the milk replacer ration must be adequate to cover the demands and provide an optimal weight gain. In Table A11.1 is a comparison of energy and protein contents between milk replacer and whole cow's milk. It provides the respective volumes and concentration of milk powder to reach the appropriate levels. However, the needs for energy and protein changes with age!

From the results in Table A11.1 one can read that – after the 3 to 4 days of colostrum – the milk replacer period starts. For the first 4 weeks give 4 meals of 1½ l (at 160 g/l) with a maximum of 2 l per meal. After that, give the calf 2 to 2½ l per meal (120 g/l) for 6 weeks, while good quality hay and starter concentrates are always provided. Weaning can then take place at 10 weeks of age. In both periods, 4 weeks and 6 weeks the demands of the calf are being met.

Table A11.1. Comparing the energy and protein contents of milk replacer with those in whole cow's milk.

Energy and protein requirements for a calf of 50 kg body weight, with a daily weight gain of 400 g	When feeding milk replacer powder (with 16 MJ; 21% crude protein/kg powder)	When feeding whole cow's milk (with 19.3 MJ; 26.4% crude protein/kg milk)
Metabolisable energy 15.6 MJ	975 g	5.8 l
Crude protein 155 g	738 g	4.2 l
Corresponds to a volume of :	8 l with 120 g powder or 6 l with 160 g powder	6 l non-diluted whole cow's milk

(after Kunz, 2008)

For calves between 2 and 20 weeks of age the volumes per meal first increase gradually, but steeply after 10 weeks of age (= weaning) from 8 to 12 l per day. At 7-8 weeks of age the volume is already at 7 l per day.

The *metabolic programming* of the calf starts around its birth. A too low energy level in the feed after birth leads to a insufficient pancreas function.

It could be that we consider *whole cow's milk cheaper* than milk replacer powder for this rearing period. In Table A11.2, a comparison is made between these two options. Whole cow's milk can be much cheaper than milk replacer powder, depending on the prevailing prices. On the other hand, one should bear in mind that cow's milk is low in e.g. Fe and Cu.

There are several kinds of milk replacer powder available on the market. There is milk replacer with milk powder, and there is milk replacer without milk powder, possibly with vegetal proteins. *Vegetal proteins* are not fit for young calves, simply because they do not have a functional rumen to digest these vegetal proteins!

Large dairy enterprises could supply 2 kinds of milk replacer to play it safely and well. For the first 4 weeks a milk replacer without, and during the next 4 to 6 weeks a milk replacer with vegetal proteins. Small dairy enterprises do not have a real choice: they have to stick to one milk replacer, one without the vegetal proteins.

It is easy to check in the field whether or not vegetal proteins are included in the milk replacer ration:
> → fill a glass with prepared milk replacer; when the colour is whitish it is
> a good milk replacer; when the colour turns yellow, it is a milk replacer
> with vegetal proteins.

Table A11.2. A comparison of the feed costs when using whole cow's milk or milk replacer powder for a period of about 10 weeks.

1 l of whole cow's milk contains 140 g DM (19.3 MJ energy + 26.4% crude protein/kg DM)	Milk replacer powder contains 16 MJ energy + 22% crude protein/kg DM
140 g DM = 2.7 MJ energy	→ equals 170 g milk replacer powder
140 g DM = 37 g crude protein	→ equals 170 g milk replacer powder
Market value = 30-40 eurocents	2.00 to 2.45 euro/kg powder
Direct costs are 20 eurocents	1.18 euro/kg powder

Let us now have a look at the *fibre content* of different protein sources (note that the level of fibre should be < 5%) as listed in Table A11.3. Table A11.4 presents an overview of different protein sources and their acceptability as calf feed stuff.

For rearing replacement heifers on a dairy farm, only option 1 is acceptable (the first column in Table A11.4) as milk replacer protein source.

In addition to *acceptability of a protein source* for the calf in relation to its fibre content, there are also other issues to address, such as the economic issue of *weight gain*.

Table A11.5 provides an overview of the daily weight gain per calf and days with diarrhoea, up to 4 weeks of age with different ration compositions of milk replacer (Kunz, 2008). The last two rations contain glycinine which a calf cannot digest.

Note that calves at 2 weeks of age have only 80% of their chymosine capacity and only 20% of their pepsin capacity available! The latter is the reason that they can not digest

Table A.11.3. Fibre contents of different protein sources.

	% crude proteins	% crude fibres
Whole cow's milk	34	--
Soja protein concentrate	67	3.6
Soja protein isolate	86	0.2
Wheat protein hydrolysate	82	0.5

Table A11.4. Protein sources and their acceptability for calf feed.

Acceptable protein source +++++	Acceptable under conditions[1]?	Marginal ----	Not acceptable ------------
Dried whey protein concentrate	Soy protein isolate	Soy flour	Meat solubles
Dried skim milk	Protein modified soy flour		Fish protein concentrate
Casein	Soy protein concentrate		Wheat flour
Dried whey	Animal plasma [2]		
Dried whey product	Animal blood cells [2]		

[1] Only acceptable when used as partial substitution for milk protein.
[2] USA conditions.

Table A11.5. Daily weight gain and days with diarrhoea in calves fed different rations of milk replacer with different protein supplements up to 4 weeks of age.

Milk replacer with…	Daily weight gain (g/day/calf)	Days with diarrhoea	Suitable as calf feed
35% skimmed milk powder + 30% lactoserum powder	617	7	+
12% soja protein isolate + 50% lactoserum powder	563	7	±
15% soja protein concentrate + 52% lactoserum powder	533	14	-
20% soja meal + 45% lactoserum powder	475	30	--

vegetal proteins (soja contains trypsine inhibitors, polysaccharides causing diarrhoea, glycinines, and beta-conglycinines with an antigenic function).

There is also the question whether or not to add *lactic acid bacteria (Cylactin®) to the milk replacer* and if so, what effect it would have on the health of the calf.

In a study reported by Kunz (2008) 4.8 mg Cylactin® was added per l of milk replacer (concentration 120 g/l) and the effect on diarrhoea and weight gain observed in calves with a birth weight of 44 kg, as compared to a group without the supplementation, up to an age of 9 weeks. The results are presented in Table A11.6.

The author did not find a statistically significant difference in weight gain between the two groups in the study (Table A11.6). With regard to diarrhoea there was a

Table A11.6. Effect of Cylactin® in milk replacer on diarrhoea and weight gain in young calves between birth and 9 weeks of age.

	Supplemented group	Comparison group
Body weight at birth	44.6 kg	43.3 kg
Body weight at 9 weeks	83.1 kg	84.0 kg
Weight gain in period	620 g/day	638 g/day
Diarrhoea cases	4	8
Diarrhoea cases treated	4	12
Mortality cases	0	0

significant difference in number of cases. This means that it could be beneficial to supplement the lactic acid bacteria product, especially in problem herds.

Instead of using milk replacer, one could consider using *whole cow's milk, but acidified*. In a study reported by Kunz (2008) formic acid (pH 4.0 to 4.6; concentration between 0.25 and 0.30%) was added to cow's milk. The number of coliform bacteria in the milk was used as outcome parameter and measured at different time intervals and temperatures:

- After 2 hrs there was no difference in coliform counts between whole milk and acidified milk at 22 °C nor at 30 °C.
- After 5 hrs, there was no difference in coliform counts between whole milk and acidified milk at 22 °C, but at 30 °C the coliform count increased to 1 million bacteria/ml whole milk while in acidified milk it stayed at a zero level.
- After 24 hrs, the whole cow's milk at both 22 °C and 30 °C had a coliform count of 2.5 million/ml milk, while the acidified milk at both 22 °C and 30 °C stayed at zero level.

This shows that, indeed, acidifying whole cow's milk could be beneficial to the calf health, reducing the pH and hampering *E. coli* infections from establishing.

Method of acidification→
- Take formic acid (pH 4.0-4.6; concentration 0.25-0.30%).
- Dilute a 85% formic acid with water in a ratio 1 to 10 (put the formic acid into the water, and not the inverse).
- *Of this solution, put 30 ml into 1 l of milk.*
- Let this combination coagulate for a while, stir it and feed it.
- This solution can be provided on the 2nd day after the colostrum period.

And the question remains: what happens to the calf health if we add *lactic acid bacteria* not to milk replacer but *to whole cow's milk*. This study too was reported by Kunz (2008).

The following feeding schedule was followed:
- 2×2 l colostrum on day 1 and 2
- 2×1½ l whole cow's milk on day 3-14
- 2×2 l whole cow's milk day 15-28
- 2×2½ l whole cow's milk day 29-56

The study group (about 90 calves) received, moreover, 6.8 mg Cylactin®/l milk. Hay and starter concentrates were supplied during the entire period.

There were no statistically significant differences found in weight gain between the two groups, not per day, not per period, not in total. The reference group showed 19 cases of diarrhoea (32 treatments) and 1 dead calf, while the study group showed 5 cases of diarrhoea and no dead calves. This difference is significant ($P<0.01$). So, acidifying whole cow's milk could be beneficial to reduce the number of diarrhoea cases.

Would this mean that whole cow's milk is the ideal feed for young calves (less costly than milk replacer, good health results with acidification, no weight gain problems)? To answer this question we need to look to the *nutrient contents of whole cow's milk too* (Table A11.7).

The conclusion is appropriate that given the results in Table A11.7, whole cow's milk is not the optimal solution to our problems! But the problem can even be worse. Of all calves born without any difficulty, about 18% have a Fe-deficiency. In calves which experienced a prolonged birth process, this percentage increases to 42%. At the same time, 33% of calves have a Hb value lower than the threshold value of 5.57 mmol Hb/l and 5% even less than 4.47 mmol Hb/l (the clinical threshold for anaemia).

Colostrum has a higher Fe level than whole cow's milk (1.16 versus 0.31 mg/l), but on average 40% of calves have no access to colostrum the first 4 hours after birth. It could, therefore, be beneficial to administer a *Fe product* to calves during the first 10 days after birth. It has been shown that weight gain was much higher in Fe supplied calves.

Other remarks

In wintertime, when ambient temperatures drop below freezing point, it is advised to supply a higher energy density milk replacer (20% fat). At the same time, the quantity of each meal must be increased by 20 to 50%. However, if additional heating in the calf barn is provided, the forenamed advice does not apply! The bags with milk replacer must be stored in a clean, dark, dry and cool place, not exposed to excessive heat. Opened bags must be stored in an air-tight manner to prevent exposure to

Table A11.7. Nutrients in whole cow's milk as compared to nutritional needs.

	Whole cow's milk	Nutrient needs (NRC, 2001)
Fe (mg/kg DM)	0.5	100
Mn (mg/kg DM)	0.3	40
Cu (mg/kg DM)	0.1-1-1	10
Co (mg/kg DM)	0.006	0.11

contamination, humidity and heat. In Table A11.8 is shown how one can assess the quality of milk replacers on the farm.

Table A11.8. Assessing the quality of milk replacers on the dairy farm.

	Features
Dry powder	
Colour	Cream to light tan, no lumps, no foreign materials.
	If powder has become orange to orange-brown, and has a caramel/burned smell, there has been a heat exposure causing loss of nutrients and loss of palatability.
Odour	Powder must have a bland to pleasant odour.
	If it smells like paint, grass, clay or petrol, the fat portion of the product may be rancid.
Reconstituted liquid	
Mixing	Product must be readily and easily soluble. Mix at recommended temperature, no clots, no deposition/sediment. Do not over-mix to avoid foam and fat separation.
Colour	Cream to light tan.
Odour	Pleasant; no off flavours.
Flavour	Milky, no off flavours.
	If organic acids are involved, the taste is often 'sweet tart'; this is not the same as the lactic acid taste of sour milk!!

Annex 12. Scoring tables for the assessment of severity and treatment of calf diarrhoea in the field (adapted after Heinrichs and Kehoe, 2009)

For a fair assessment of the diarrhoea situation among young stock, three scoring sheets are handled: (1) on the extent of diarrhoea, (2) on respiration of the calves, and (3) on appearance of the calves.

Table A12.1 provides a scoring sheet for assessing diarrhoea cases in young calves; Table A12.2 provides a scoring sheet for assessing the quality and its deviations of respiration; Table A12.3 gives a scoring sheet for assessing clinical appearance of the calves (Heinrichs and Kehoe, 2009).

On the basis of the outcome of all scoring tables, a *daily total* is calculated. When the daily total exceeds a certain threshold (for example an overall average of 8), rectal temperature must be taken, and treatment started (antibiotics and or oral rehydration).

Table A12.1. Scoring system for distinguishing different degrees of diarrhoea in young calves.

Score 1	Calves show normal faeces, with *'pudding'* consistency
Score 2	Calves show slightly less firm faeces which are *'yoghurt-like'*
Score 3	Calves show diarrhoea, with *'syrup-like'* faeces, which are loose to watery; odour
Score 4	Calves show diarrhoea with *' fruit juice'* aspect; faecal material still visible
Score 5	Calves show *'watery'* diarrhoea without faecal material, with mucous or blood

Table A12.2. Scoring system for distinguishing different degrees of respiration/deviations.

Score 1	Calves show normal breathing, no clinical signs what-so-ever
Score 2	Calves may show a slight cough or runny nose; but respiration is regular
Score 3	Calves show rapid breathing and moderate coughing
Score 4	Calves show severe frequent cough, and rapid breathing
Score 5	Calves show irregular breathing, with severe chronic cough

Table A12.3. Scoring sheet for assessing the clinical appearance of young calves.

Score 1	Calves are alert and active
Score 2	Calves have droopy ears, and are slightly unresponsive to stimuli
Score 3	Calves are moderately depressed, with head and ears drooping
Score 4	Calves are depressed, with drooping head and ears; show no interest in standing up
Score 5	Calves are down all the time (side position)

Note that from a score of 2 onwards (for each Table) one can already speak of a disorder. The summing up of 3 scores at value 2 is 3×2= 6. So, a very strict threshold would be an overall average daily score of 6.

Example of calculation:
- 15 calves were scored, using the three Tables A, B and C
- Results of scoring are A (10×3+5×2)+B (9×2+6×3)+C (6×2+9×4) = 124
- The overall average is 124/15 calves = 8.3
- If the threshold was set at 8, rectal temperature must be taken and treatment of calves started

Section II. Diseases of young stock

(with special emphasis on diseases affecting groups of calves, e.g. IBR and BVD)

Chapter 8. Introduction to Section II

Calves are the future of the dairy herd. A minority of the female animals will be kept on the farm as herd replacements while male calves are sold, except on farms which are producing males for reproduction or beef production purposes.

Section II does not provide the reader with text-book and detailed descriptions of all diseases, their diagnosis, treatment and prevention. Textbooks which cover these areas more comprehensively are available. The objective is rather to provide the reader with practical and easy-to-consult guidelines on various major diseases and disorders in calves, affecting the group as a whole rather than individuals. Such diseases and disorders affect the profitability of the young stock rearing process and hence the dairy farm. The aim is to help veterinarians (and extension officers or farmers) to make better decisions, and help them to increase the profitability of the young stock rearing process.

Presented here are the name of the disease and its occurrence, its clinical signs, therapeutic (medical and management) measures, prophylactic possibilities and, if necessary, how to explain to the farmer the problem or problems that he could have to deal with, when trying to eliminate or reduce the diseases present on his farm. The selected diseases and disorders are those with the greatest impact on herd health and productivity which require FHPM programmes to control them rather than diseases with a low within-herd prevalence affecting only individual animals. They cause either high mortality or morbidity rates, or in some cases, both (Table 8.1).

The perinatal period is the period of greatest young stock mortality in dairy herds. Enteric infections are the primary cause of mortality in young calves. Colibacillosis is the number one cause of enteric mortality in neonatal calves. BVD is economically the most important endemic viral disease of cattle worldwide, whereas Johne's disease is one of the most economically important bacterial disease of cattle and potentially related to Crohne's disease in humans. IBR is one of the most important diseases affecting national and international trade of cattle and semen and hence the AI industry globally. The other diseases selected are those which present a serious risk to young stock through either their high prevalence or the economic consequences of an outbreak in a dairy herd, or their geographical interest.

In the following paragraphs, the major causes of calf disease and mortality will be addressed. Then, special attention is given to issues of IBR and BVD, followed by other major diseases and tropical/subtropical diseases in a summary style. The focus is on disease, including prevention, in a practical way so it can be useful in the farmer's daily work. No in-depth explanations of complex pathogenesis of diseases will be

Table 8.1. The most common causes of mortality, by age group, of young stock. Ranking is in descending order of importance.

From birth to 1 month of age	From 1 to 3 months of age	From 3 to 12 months of age	From 12 months to 1st calving[1]
Enteric infections	Respiratory infections	Respiratory infections	Respiratory infections
Septicaemia/Bacteraemia (after e.g. enteric or respiratory infection, or navel joint illness)	Enteric infections	Clostridial infections	Digestive disorders
Respiratory infections	Septicaemia/ Bacteraemia	Enteric infections	Dystocia
Navel joint illness	Gastrointestinal obstruction or torsion	Septicaemia/ Bacteraemia	Other infections
Gastrointestinal obstruction or torsion	Abomasal ulceration	BVD	Poisoning

[1] Note that the primary reason for heifers in this age cohort failing to enter the milking herd is commonly culling for infertility (freemartins and failure to become pregnant), not on-farm mortality. (Brickell *et al.*, 2009, RVL, 2008, Svensson *et al.*, 2006)

given; emphasis will be on occurrence, diagnosis, treatment, prevention and overall management.

Definitions

In the context of this book, the following terminology will be used:
- The *transition period* refers to the three weeks before and after calving and usually deals with problems of the dam.
- The *perinatal period* refers to the time period from 24 hrs prior to birth, the birth process itself, to 48 hrs after birth.
- An *abortion* refers to a calf or calves which have died and are delivered by the dam between 100 and 260 days of gestation.
- A *stillbirth* refers to a calf or calves (twins or more) which are dead at birth (die before or during calving) following a gestation period of at least 260 days (excludes abortions).
- The *neonatal period* refers to the calf from 3 to 7 days of age.
- *Young stock* refers to parity 0 animals up to first calving.

Chapter 9. Perinatal disorders

9.1. Introduction

The perinatal period is the most hazardous in the life of all animals. The main causes of perinatal morbidity and mortality are, in descending order of importance, combined respiratory and metabolic acidosis, parturient trauma, hypoglobulinaemia, congenital infections and deficiencies and omphalophlebitis. *Perinatal mortality* may be defined as calf death prior to, during or within 48 hours of calving, following a gestation period of at least 260 days, irrespective of the cause of death or the circumstances of the calving (Mee, 2009). Perinatal morbidity and mortality are growing welfare concerns, given their impact not just on losses around calving but also on subsequent productivity, health, reproduction and farm economics. In this context it is critically important to pay sufficient attention to the beginning of the calf's life, because any mistake made there will lead to problems, economic loss and stress in the farmer.

Management of the newborn dairy calf is best achieved through implementation of simple protocols which document the correct strategies to be followed at the herd level and the correct procedures to be carried out at the individual animal level. These protocols cover management of calving (monitoring of eutocia and detection and management of dystocia) and newborn calf care. Discussion with producers about newborn calf problems or care represents a contact moment which veterinarians should utilize to expand their role in veterinarian-led dairy herd management support programmes (Mee, 2007). See also the Annexes.

For an analysis of the incidence of any disease, it is necessary that the farmers record such data continuously and properly (Noordhuizen *et al.*, 1983). At herd level, laboratory analysis will be useful in detecting subclinical diseases and risk factors (Gelfert and Staufenbiel, 2004), which may increase the disease incidence; such data too must be recorded.

The data recording procedure must be as easy as possible to avoid false entries in the data base (Krebs *et al.*, 1999). Therefore, before starting such recording, diseases have to be defined clearly (Kelton *et al.*, 1998). Examples of practical recording systems for calf rearing have been presented by Brand *et al.* (1996) and can be found in Section I too.

9.2. Management of the dairy cow at calving to prevent losses in young stock

9.2.1. Dry cow body condition score

During the transition period particular attention must be given to the body condition score (BCS) of pregnant heifers and cows. Over-conditioning must be avoided, because this can induce fatty liver disease, sub-acute ketosis (with high prevalence of *postpartum* diseases in cows, a poor colostrum quality, and increased health problems in calves). Other potential consequences of over-conditioning are the occurrence of more dystocia and a delayed birth, with acute asphyxia in the neonate or late asphyxia. Ideally cows should calve down at a BCS of 3.25 (range 3.0-3.5, on a scale of 1-5). In order to achieve this score, BCS data need to be recorded in late lactation and shortly before drying off. In case of deviations, the rations for late lactation cows must be adjusted, and ultimately the length of the dry period if no other means are at disposal (Cannas da Silva *et al.*, 2006).

9.2.2. Movement to the maternity unit

In order to avoid increased calving problems and perinatal mortality associated with cows calving in the dry cow accommodation, producers should ensure cows calve in a maternity unit. Within 36 hours of calving, cows in cubicle housing will attempt to seek isolation by lying in cubicles furthest from other cows, thus early movement to the maternity unit fulfils their natural isolation seeking behaviour. It is recommended that pregnant cows are in the maternity unit at least 24 hours before calving and heifers earlier, as this is when calving commences. Movement too early to the maternity unit will affect the cleanliness of the calving environment and increase the risk of ketosis and displaced abomasum. The movement of animals, particularly nervous cows and heifers, will suspend their calving behaviour so they should be left without further disturbance to adapt to the maternity unit and resume calving progress. Environmental stress can be further reduced in heifers by calving them separate from older cows, maintaining visual contact with herd mates to prevent social isolation, not tethering them at calving and avoiding disturbances from routine farm tasks, such as calf feeding and cleaning out pens.

An alternative strategy is to move cows once stage two (commenced calving) has been detected. The advantages of this approach are that cows spend hours, not days, in the maternity pen thus reducing the magnitude of the change in dry matter intake and freeing up maternity pens for other cows, particularly in large herds. Recent research indicate that moving animals in stage two is associated with reduced time to first lie down, duration of calving from entry and reduced assistance, dystocia and stillbirth rates. These results suggest that it is less detrimental to move animals which have already commenced calving (stage two) than it is to move animals which are about to start calving (stage one). However, this strategy requires 24 hour monitoring of the close-up group with approximately hourly checks and it is not clear whether this policy may interrupt the calving process and lead to more calving problems than if these animals were not moved or were moved before stage one commenced.

Currently, moving cows before calving commences, as is widely practised, appears prudent to optimise newborn calf care. However, the potential management benefits of alternative strategies, particularly in large dairies, need to be considered by farm managers (Mee, 2008a).

A system becoming more and more adopted on larger modern dairy farms is the principle of housing and feeding dry cows in one group in large straw pens. Benefits of this system are:
1. cows are housed in the herd during dry period but have enough space to separate themselves at calving;
2. dry cows in late gestation get up more easy on straw compared to the cubicle situation;
3. no ration changes prior to calving;
4. no need to determine when to move the cow prior to calving, therefore the risk of moving cows in stage one of calving and no detrimental effects of stress due to this action.

Prerequisites of this system are that fresh straw should be applied at least once daily and calves should be removed immediately after calving (which is not different from any other system).

9.2.3. Calving supervision

Good supervision is dependent upon monitoring calving, particularly stage two, and intervening if and where necessary, while avoiding excessive direct supervision. Monitoring approximately every 3 to 6 hours from the first detection of the onset of stage one is advisable to detect the onset of stage two of normal calving and to detect abnormal calvings early. If cervical dilation has commenced and no abnormalities are detectable monitoring should continue approximately hourly. The allantochorion ruptures approximately an hour before the amnion appears. Undue delay between the rupture of the allantochorion and the appearance of the amnion or fetal hooves may indicate a problem such as fetal oversize or maldisposition (Mee, 2008b).

9.2.4. Calving assistance

The vast majority of cows will happily calve unattended and unassisted and where possible should be allowed to do so. However, a small proportion of cows and a greater proportion of heifers may require assistance. Phenotypic dystocia rates are increasing internationally and currently vary between 2 and 15% (Mee, 2008c). Three simple questions need to be addressed by herd personnel to ensure successful calving intervention: (1) whether or when to intervene, (2) how to intervene and (3) when to solicit veterinary assistance. For the veterinarian these queries often come down to whether to pursue traction or surgery. Potential dystocia may be differentiated from eutocia by the presence of risk factors for the various types of dystocia. These risk factors may be assessed from the calf sire, breed and size and body condition and size of the dam, previous calving history, exploratory examination, calving conditions and most importantly, calving progress. Addressing the question of whether to intervene during calving, intervention is recommended in cases of feto-pelvic incompatibility (FPI), maldisposition, twinning, uterine inertia and vulval or cervical stenosis. Addressing the question of when to intervene, early intervention is recommended, during stage one for uterine inertia, and during stage two, for maldisposition and twinning. Delayed intervention is recommended, during stage two, for FPI and cases of vulval or cervical stenosis. FPI with a live full-term normal calf in anterior presentation is the primary reason for intervention during calving, particularly by herd personnel. The importance of progress, rather than clock-watching during stage two is emphasised, as the onset of stage two is usually unknown. Signs of progress during stage two include a recumbent dam straining intermittently but strongly, with occasional breaks while she stands up and lies down again and progressive emergence of the fetal legs and head through the vulva. Once progress is normal, discrete monitoring without disturbance every 30 minutes, or continuously if patience can be assured, is recommended. Intervention should not be carried out before the calf's muzzle has emerged and not before the calf's fetlocks are visible. When progress ceases over 30 minutes or the calf begins to exhibit signs of reduced vigor (such as capital or lingual oedema, buccal or lingual cyanosis, scleral haemorrhages intervention should be conducted. When severe acidosis can be traced back to stage two of relatively short duration, rapid improvement can be achieved by resuscitative care. When acidosis exists over a longer period, as in delayed assistance, the efficacy of supportive care is lower as hypoxic lesions such as meningeal, subepicardial and subpleural haemorrhages may develop. The stress of a prolonged delivery, rather than the type of assistance is responsible for reduced calf vigor following dystocia.

9.3. Perinatal mortality

The prevalence of perinatal mortality in dairy herds internationally has increased in recent years and currently varies between 2 and 10% (Mee *et al.*, 2008; Heinrichs and Kehoe, 2009). This average figure obscures the fact that perinatal mortality follows a right skewed distribution where most herds have none or minimal losses but some herds have high (25%) mortality. Some 90% of calves, which die in the perinatal period, were alive at the start of calving and so much of this loss is preventable. Recently an increase in perinatal mortality in heifers following non-apparent dystocia has become apparent (Mee, 2008d) which may have a genetic basis. Significant herd and animal-level risk factors are shown in Table 9.1.

Traditionally, the majority of perinatal mortality has been attributed to dystocia. Recent research indicates that the proportion of perinatal mortality attributable to dystocia may be decreasing. The main causes of perinatal mortality are anoxia (*asphyxia neonatorum*) and trauma, following dystocia, and to a much lesser extent death-in-utero and premature placental expulsion. Perinatal mortality following eutocia (often called 'weak calf syndrome') may be associated with IntraUterine Growth Retardation (IUGR) or prematurity, congenital defects, infections, precalving nutrition, dysmaturity, twins, placental dysfunction or sire-specific genetic weakness leading to poor perinatal viability, prolonged stage one with premature placental separation or prolonged stage two with uterine atony. A comprehensive list of possible causes of perinatal mortality is given in Table 9.2.

9.4. Asphyxia neonatorum

Perinatal asphyxia is complex disease, provoked by a respiratory metabolic acidosis, which is easy to diagnose clinically because the calf shows respiratory difficulties and a low vitality. Asphyxia immediately after birth ('acute asphyxia') or 'late asphyxia'

Table 9.1. Herd and animal-level risk factors for perinatal calf mortality.

Herd level	Individual animal level
Calving supervision	Parity of the dam
Calving environment	Dystocia
Nutrition *prepartum*	Gestation length
Congenital infections	Birth weight of calf
Herd size?	Gender of calf
(after Mee, 2008e)	

Table 9.2. Infectious and non-infectious causes of stillbirth and perinatal death in calves.[1]

Non-infectious causes	Infectious causes 1	Infectious causes 2	Congenital phenomena
Dystocia	**Virus**	**Bacteria**	Epitheliogenesis imperfecta
Cold stress	BHV-I	*Brucella abortus*	Cardiac defects
Excess or under-nutrition	BVDV	*Brucella millitensis*	Internal hydrocephalus
Energy deficiency	Bluetongue virus	*C. pyogenes*	Cerebellar hypoplasia
Protein deficiency	Akabane virus	*Campylobacter* spp.	Arthogryposis or cleft palate
Ketosis	Cache valley virus	*Salmonella* spp.	Beta-mannodiosis
Umbilical haemorrhage		*Bacillus* spp.	
Copper excess/deficiency	**Haemetoza**	*Streptococus* spp.	Gama-mannodiosis
Selenium and iodine deficiency	Anaplasmosis	*Leptospira* spp.	Bovine citrullinemia
Vitamin A deficiency	Theileriosis	*Listeria monocytogenes*	Bovine maple syrup urine disease
Premature placental separation	**Fungus**	Navel infection	Congenital joint laxity and dwarfism
Intrauterine growth retardation	*Aspergillus* spp.	**Protoza**	Intestinal atresia
Nitrate toxicity	**Rickettsia**	*Neospora caninum*	Omphalocoele
Twinning	*Clamydia* spp.	*Toxoplasma gondii*	Anophthalmia
Uterine torsion	*Coxiella burnetii*	*Tritrichomonas foetus*	Prosencephalic hypoplasia

[1] Note that the relevance of each cause may differ between countries and/or regions.
(adapted after Smith, 2009)

can be distinguished. Both asphyxia types lead to a high rate of perinatal deaths. Acute asphyxia develops inside the uterus namely if the cow has a uterine torsion, the presentation of the calf causes the disease, the uterine contractions are weak, or when the intervention at calving is too late. The calf suffers from an oxygen deficiency and shows a high level of CO_2 due to a reduced gas exchange between placenta and foetus; the calf will suffer from a metabolic acidosis which it tries to compensate by respiration. Two scenarios are possible: (a) intra-uterine death, or (b) neonatal asphyxia (Grunert *et al.*, 1984). This situation can provoke a high rate of calf losses. Following successful perinatal resuscitation many calves that experience dystocia or prolonged calving will still have a mixed metabolic-respiratory acidosis in the first six hours of life (late asphyxia). In addition, some calves develop secondary acidosis within 24 hours of birth with a poor suck reflex, tachypnoea, tachycardia, weakness, depression and hypothermia. Such late perinatal acidosis is significant risk factors for perinatal mortality. Correction of postnatal metabolic acidosis can be effectively achieved with drip or bolus intravenous infusion of sodium bicarbonate instituted after resuscitation and repeated as necessary.

9.5. Perinatal calf evaluation

All calves suffer some degree of respiratory-metabolic acidosis at birth, but calves born following prolonged calving have increased respiratory and metabolic acidosis. Calves assisted compared to unassisted or pulled out by strong compared to mild traction have increased respiratory-metabolic acidosis and take longer to achieve sternal recumbence. The vigour of the calf can be assessed immediately after calving by its reflexes and the time it takes to head-right, achieve sternal recumbence, attempt to stand and to stand (3, 5, 20 and 60 minutes, respectively). If the calf exhibits superficial abdominal breathing, has poor reflexes or it takes more than 15 minutes to achieve sternal recumbence the prognosis is poor (Schuijt and Taverne, 1994).

9.6. Calf resuscitation

Most calves which require resuscitation are usually not attended by a veterinary practitioner as most dystocia and prolonged calvings are attended by herd staff only or are unattended. Hence, the role of the veterinary practitioner in calf resuscitation is two-fold. Firstly to draw up a standard operational procedure, SOP, (a technical and formalised work instruction) for at-risk calves for herd staff and secondly to resuscitate calves after veterinary-assisted calvings.

The SOP should document for herd staff a standard resuscitation kit to be located in the maternity pen area and details of first-aid procedures to be followed with at-risk calves. At-risk calves are those who are likely to need resuscitation because of their calving outcomes. Before starting resuscitation one should make sure that the upper respiratory tracts are free from fluid, mucus and attached fetal membranes. Preferably this is done by elevating the rear of the calf. When a calf in up-side-down position does not start breathing immediately in response to brisk rubbing on the chest, resuscitation is needed. A resuscitation kit for herd staff on large dairies should include a stethoscope, rectal thermometer, compressed air device (e.g. Ambu bag), needles, suction pump and oxygen delivery equipment. For successful resuscitation of at-risk calves, herd staff need to practice good calving supervision (i.e. be present to assist the calf), prompt calf viability evaluation (during and immediately after birth) and early aggressive intervention (i.e. active management of calving and calf care). Note that newborn calves are not likely to survive if resuscitation does not result in spontaneous respiration in 2 or 3 minutes (Noakes *et al.*, 2001).

9.7. Umbilical care

After spontaneous rupture of the umbilical cord, the urachus and vessels normally retract into the abdomen thus protecting them from environmental contamination. At assisted calvings there is a tendency to immediately rupture the cord. Though research in calves is limited, one study found a long-term decrease in efficiency of pulmonary gas exchange in calves with assisted premature umbilical cord rupture compared to those with spontaneous rupture. Thus, there may be an advantage to leaving the cord to rupture spontaneously.

Omphalitis or navel illness occurs in 5 to 15% of newborn calves. As it is generally untreated, it can lead to reduced growth, joint ill and other sequelae. Prevention of navel illness is based on maintenance of maternity pen hygiene, reducing the residency time of calves in unhygienic maternity pens, ensuring adequate early intake of good quality colostrum and navel antisepsis. Choice of cord care procedure is under-researched in calves. Topical antiseptics, particularly iodine and chlorhexidine (dip or spray), are more widely used in calves than topical antibiotic spray or cord clamping or ligation. In herds without umbilical-associated problems, farmers should avoid possibly harmful cord application procedures and concentrate on maternity hygiene and calf immunity. In herds with serious navel-ill problems producers should consider improving maternity pen hygiene, immediate and repeated cord dipping with chlorhexidine or iodine, calf snatching, hand-feeding colostrum and regular checking for omphalitis (swollen and painful navel, pyrexia) with metaphylactic parenteral antimicrobial therapy as appropriate.

9.8. Calf movement after calving

In herds where paratuberculosis may be present based on a risk assessment, clinical history or laboratory results, newborn heifer calves in particular should not be allowed to suckle. After immediately removal from the dam ('calf snatch') and placement in a calf house or hutch, they should be fed hand or machine-milked colostrum from their dam. Where the risk of infectious disease is acceptably low calves may benefit from remaining with their dam to increase their opportunities to suckle naturally and to enhance the absorption of colostral immunoglobulin over colostrum fed in the absence of the dam. As cows tend to lick off antiseptics applied to the umbilicus, such antiseptics should be reapplied upon removing the calf from the maternity pen. Where maternity pen hygiene is poor, calf residency time should be reduced to a minimum to prevent common calf-hood infectious disorders such as omphalophlebitis, diarrhoea, pneumonia and septicaemia. Moving the calf from the maternity pen to the calf house presents an opportunity to conduct a quick check on the calf's health status. Problems to look out for include persistent signs of acidosis, dyspnoea, umbilical bleeding or organ eventration and hypothermia.

9.9. In conclusion

Despite advances in dairy herd health and productivity management support, perinatal calf mortality rates are still unacceptably high on many dairy farms. Whilst some of this loss has a genetic origin and may be outside the producer's control, FHPM programmes at the herd level and SOPs at the animal or group level can be implemented to improve perinatal welfare. The key features of successful newborn dairy calf management are ensuring heifers and cows are moved in time to calve in suitable maternity housing, discrete calving supervision and appropriate timing of any necessary calving assistance, immediate parturient evaluation of at-risk newborn calves followed by aggressive resuscitation, strategic navel antisepsis, early detection (and treatment) of perinatal problems and prompt movement of the newborn calf to hygienic calf housing. Veterinarian-led producer implementation of FHPM programmes in the management of calving and newborn calf care can improve young stock welfare and health. A relevant component of such a programme is the recording and evaluation of the various events in the perinatal period in order to improve overall performance of farm management and farm employees. The relatively high incidence of calf mortality warrants a structured and continuous monitoring of calf performance and an advisor – coaching role of the veterinarian.

Chapter 10. Infectious calf diarrhoea and septicaemia[1]

10.1. Introduction (principal agents of diarrhoea in calves)

Diarrhoea in the new-born calf is an important cause of calf mortality and hence economic loss. The primary, new-born calf's diarrhoea agents are *E. coli*, Rotavirus, Coronavirus, *Cryptospidium parvum*, *Eimeria* spp., and *Salmonella* spp. But *C. parvum* has become increasingly important in the last couple of years (Martins *et al.*, 2007a,b).

Besides the agents named, there are other important, contributory risk factors such as: (a) poor management, (b) inappropriate environmental conditions, (c) deficient administration of the colostrum (given too late, in wrong quantities or at poor quality), (d) influence of feeding in the dry period of the dam (low level of antibodies), (e) the inexistence of prophylactic schemes (Radostits, 2000).

Although there are different infectious agents causing neonatal diarrhoea (0-7 days old) (Table 10.1), *E. coli* is still the most important one. Relevance of other agents may differ between husbandry systems.

Table 10.1. Age occurrence of diarrhoea in calves.

Agent	Bacteria	Virus	Parasite	Zoonosis	Age (days)
Neonatal					
Septicaemia by *E. coli*	X				<2
Enterotoxigenic *E. coli* (ETEC)	X				0-7
Enterohaemorrhagic *E. coli* (EHEC or VTEC)	X			X	0-7
Scours					
Rotavirus		X			5-15
Corona virus		X			5-21
Cryptosporidium parvum			X	X	5-35
Salmonella spp.	X			X	5-42
Clostridium perfringens types B and C	X				5-15
Eimeria spp.			X		>30

(Radostits, 2000 and 2007, Smith, 2009)

---

[1] The text in this chapter is derived from different publications and oral presentations given by Radostitis (2000), Radostits *et al.* (2001a,b), Radostits *et al.* (2007), Smith (2009).

10.2. Infectious calf diarrhoea and septicaemia

10.2.1. Synergism between enteropathogens

The synergism between Rotavirus and enterotoxigenic *E. coli* in calves older than 2 days may explain the fatal diarrhoea which can occur in calves at 1 week of age. In short, septicemic colibacillosis occurs in newborn animals which are agammaglobulinemic because they have not ingested sufficient colostrum early enough, or absorbed sufficient colostrum but with too low immunoglobulin content. Enteric colibacillosis occurs in colostrum-fed animals and is caused by the colonisation and proliferation of enteropathogenic *E. coli*. For these infections, the primary infection with either Rotavirus or *Cryptosporidium* spp. or their combination, functions as port of entrance. Rotavirus destroys the top of the villi, enhancing the colonisation by *E. coli*.

During severe disease outbreaks, it is often necessary to conduct a necropsy on diarrhoeic animals which have been killed specifically for the purpose of obtaining a definitive aetiologic diagnosis. The combined use of bacteriological, parasitological and virological methods, together with histological and immuno-fluorescent studies of fresh intestinal tissue will provide the most useful information about the location of the lesions and the presence of entero-pathogens. *Post-mortem* autolysis of the intestinal mucosae and invasion of the tissues by intestinal micro-flora occurs within minutes after death, so gut or rectum samples should be collected immediately following euthanasia of the animal.

Calf-side tests are also available to have test results 'within minutes' and to know whether *E. coli*, Rotavirus, *Cryptosporidium* spp., *salmonellae*, *Clostridium* spp. or *Eimeria* spp. are present. These tests are cheap and reduce the interval between sampling and test result; moreover, they can be used to show the farmer what is the likely cause of the calf diarrhoea in his herd.

Rotavirus

- *Occurrence*: as single case or as mass infection in the first weeks of life.
- *Signs*: loss of appetite, tenesmus, profuse bubbly diarrhoea with mucosa particles. Further, severe complications occur when also *C. parvum* is present as co-infection.
- *Diagnosis*: ELISA, PCR. Calf-side tests are available.
- *Prevention*: through vaccination of dam 2 months prior to calving date, and the provision of sufficient quantity and high quality colostrum to the calf after birth. High hygiene standards must be achieved, see also risk factors and prophylaxis further down.

- *Treatment*: electrolyte solutions intravenously and orally. Antimicrobials to avoid secondary bacteriological infections.

Corona virus

- *Occurrence*: as single case or as mass infection. A massive destruction of epithelial cells of the mucosa of the intestines occurs, with loss of villi.
- *Signs*: loss of appetite, tenesmus, profuse bubbly diarrhoea with mucosa particles. From signs it can not be distinguished from Rota-infections; often in combination with Rotavirus
- *Diagnosis*: ELISA, PCR. Calf-side tests are available.
- *Prevention*: through vaccination of the dam 2 months prior to calving date, and the provision of sufficient quantities and high quality colostrum to the calf after birth. High hygiene standards must be achieved, see also risk factors and prophylaxis further down.
- *Treatment*: electrolyte solutions intravenously and orally. Antimicrobials to avoid secondary bacteriological infections.

Cryptosporidiosis: C. parvum

- *Occurrence*: as single case or as mass infection in the first weeks of life. Is associated with Rota-, Corona virus and *E. coli.*
- *Diagnosis*: ELISA, PCR ELISA, Immunofluorescence, Oocyst detection by flottation.
- Prevention: separation of calves from dams; high hygiene standards; identification and elimination of carrier cows. High hygiene standards must be achieved, see also risk factors and prophylaxis further down.
- *Treatment*: halofuginon lactate (most effective until about 7 days of age). Electrolyte solutions intravenously and orally. Antimicrobials to avoid secondary bacteriological infections.

*Salmonellosis (*S. typhimurium; S. dublin*)*

- *Occurrence*: between 2 and 26 weeks of age most frequently.
- *Signs*: fever, anorexia, yellow diarrhoea, brownish, fetid, cough, arthritis, respiratory problems, septicaemia.
- *Diagnosis*: faecal samples (but intermittent shedding can occur), blood (paired samples 3 weeks apart for seroconversion), aborted fetus (for *S. Dublin*-induced abortion).
- *Prevention*: vaccination is a means to prevent the disease, if allowed.
- *Treatment*: with sulfamethazin, or other appropriate antibiotics.

- *Prophylaxis*: could be through strict hygiene (and possibly vaccination?).
- The disease must be notified to the proper authorities.

Clostridiosis *(Cl. perfringens types a, b, c, d, e; Cl. sordelli)*

- *Occurrence*: types B and C are uncommon infections in calves at the age of 7-10 days up to 10 weeks of life. Type D:Veal calves between 1 and 4 months of age.
- *Signs*: colic, frequent diarrhoea with blood, arthritis, bloat, opisthotonus, commonly a rapid death.
- *Diagnosis*: clinical signs, necropsy findings and bacteriology.
- *Treatment* of *Cl. perfringens* or *Cl. sordelli* often comes too late; one could try the betalactamates.
- The disease is not very contagious, and often endemic in a dairy farm.

Eimeria *spp.*

- For *Eimeria* spp. see Chapter 5 of Section II.

Nutritional diarrhoea

Risk factors:
- Wrong temperature of colostrum or milk replacer when given to calf.
- Too high quantity of colostrum given in one meal.
- Irregular or incorrect feeding interval between colostrum or milk replacer meals.
- Poor quality or type/composition of milk replacer given the age of the calves (see Annex 8).
- Poor composition of feedstuffs and management failures at feeding.

10.3. Colibacillosis

10.3.1. E. coli *infections*

Infection with pathogenic serotypes of *E coli* is called colibacillosis. There are two types of colibacillosis; septicaemic and enteric. These infections affect new-born calves and the prevalence of enteropathogenic *E. coli* varies between herds. There are 4 broad categories of enteropathogenic *E. coli* infections: septicaemic, enterotoxigenic, enterohaemorrhagic and necrotoxigenic.

The most common enteropathogens which cause diarrhoea in neonatal farm animals are the *enterotoxigenic E. coli (ETEC)* which are not invasive and cause hypersecretion by adhering, colonising, and producing enterotoxins in the small intestine. *E. coli* infections in the first 2 days of life often cause enteric septicaemia, often followed by post-sepsis diseases like (poly-)arthritis.

Enterohaemorrhagic E. coli *(EHEC)* is an uncommon cause of disease of newborn farm animals and attach to the colon and distal small intestine, '*attaching and effacing* E. coli *(AAEC)*', resulting in a haemorrhagic colitis. They are also known as *verocytotoxic* E. coli *(VTEC)* because they produce verocytotoxins. Necrotoxigenic E. coli *(NTEC)* lead to necrotic diarrhoea (after Radostits, 2000).

Risk factors for colibacillosis include for example: (a) colostrum deprivation or poor quality colostrum, (b) overcrowding, (c) adverse climatic conditions, (d) poor quality milk replacers or poor milk replacer management, (e) poor hygienic conditions.

Zoonosis: certain serotypes of enterohaemorrhagic E. coli found in animals are pathogenic for humans and the zoonotic implications must be considered when dealing with coliform infections in farm animals (Table 10.2). Cattle are an important but limited reservoir for E. coli serotype $O_{157}H_7$ (prevalence about 12%) which causes haemorrhagic colitis and the haemolytic uremic syndrome in humans (Radostits, 2001; Smith, 2009).

Table 10.2. Enteropathogenic E. coli *infections (colibacillosis) of calves.*

E. coli serotypes	Primary risk factors	Age group affected	Mechanisms of action	Clinical signs	Zoonosis risk
Septicaemic	Insufficient colostral IgG absorption	0-4 d	Invasive, endotoxins	Septicaemia, diarrhoea	No
Enterotoxigenic	ETEC infected cows and environment	0-4 d	Adhesion, colonisation, enterotoxins	Diarrhoea and shock	No
Enterohaemorrhagic	EHEC infected cows and environment	2-5 wks	Shiga and verocytotoxins	Haemorrhagic colitis	Yes
Necrotoxigenic	NTEC infected cows and environment	<3 mths	Necrotoxins	Necrotic diarrhoea	No

Septicaemic E. coli

- *Disease type*: Enteric septicaemia
- *Occurrence*: in newborn calves. May occur as early as 12 to 18 hours after birth; occasionally in calves up to several days of age - mixed infection with viral enteropathogens. Often seasonal pattern.
- An adequate level of serum immunoglobulins, IgG, protects calves from death.
- *Risk factors*:
  - Age/parity of dam; the risk of diarrhoea in calves born from heifers may be about 4 times greater than in calves born to cows because conformation of the udder of heifers may hamper suckling. Poor mothering due to mistaken identity may occur, resulting in the calf not receiving any or adequate colostrum.
  - Weight at birth and birth process; calves weak or having an oedematous tongue after a dystocia, may not be able to suck properly for several hours.
  - Calf pen conditions; cold, wet, windy weather during the winter months and hot dry weather during the summer months have a significant effect on dairy calf mortality. Overcrowding is commonly followed by an outbreak of acute diarrhoea in calves. Housing and hygiene practices are probably the most important risk factors influencing the incidence of septicaemic colibacillosis in calves.
  - Colostrum management; newborn farm animals could be agammaglobulinemic if they do not ingest a sufficient amount of high quality colostrum soon enough after birth.
  - Farm and feeding management; the person feeding and caring for the calves is an important factor influencing calf morbidity and mortality due to diarrhoea or other diseases. Heifers are commonly more closely confined during the calving season for more effective observation and assistance at parturition; this may lead to increased contamination of the environment.
  - Simultaneous enteric infections; Rotavirus and *Cryptosporidium* spp. may enhance colonisation of the *E. coli*.
- *Signs*:
  - As early as 12 to 18 hours after birth; diarrhoea; dehydration; acidosis; death.
  - Arthritis is a common sequel in calves as well as meningitis. If fluids are lost extensively, hypovolemia and shock occur, as well as lactic acidosis, hypothermia, tachycardia, weak, recumbent, capillary refill time may be prolonged, cold extremities, weak peripheral pulse.
  - The illness is peracute, the course can vary from 24 to 96 hours with a survival rate of less than 12%. There are no clinical signs of strict diagnostic value.
  - If a calf survives the septicaemia state, clinical evidence of post-sepsis localisation may appear in about 1 week. The clinical signs of such situation are: arthritis, meningitis, panophthalmy, pneumonia, opisthotonus, convulsions, tremors.

- Necropsy findings
  - Findings: No gross lesions (diagnosis may depend upon isolation of organism from filtering organs). In less severe cases: may present 1-subserosal, 2-submucosal haemorrhages. Enteritis and gastritis may be present. Fibrinous exudates occasionally found in joints and serous cavities. Possible omphalophlebitis, pneumonia and meningitis.
  - Samples for confirmation of diagnosis:
    ▷ Bacteriology: spleen, lung, liver, culture swabs of exudates, umbilicus, meninges.
    ▷ Histology: formalin-fixed samples of spleen, lung, liver, kidney, brain and any gross lesions.
- *Differential diagnosis*:
  - *Salmonella* spp.; *Listeria monocytogenes*; *M. haemolytica*; *Streptococcus* spp.; *Pneumococcus*; *Cryptosporidium parvum*.
- *Treatment*:
  - Oral re-hydration; fluid therapy intravenously.
  - Oral antibiotics and parenteral antibiotics.
- *Prophylaxis*:
  - Ensure the calf ingests sufficient and high quality colostrum and absorb colostral immunoglobulins within minutes to 1 hour after birth to obtain proper protection against septicaemia and enteric colibacillosis.
  - Natural sucking by the calf may possibly enhance the efficiency of absorption of colostral immunoglobulins, but should not be the preferred practice because control on intake is lacking! For biosecurity reasons calves should be separated from their dam immediately after birth (paratuberculosis, see Paragraph 6.7). High hygiene standards must be achieved.

Enterotoxigenic E. coli

- *Disease type*: Enterotoxigenic colibacillosis
- *Occurrence*: at 3 to 5 days of age is the most common form.
- *Risk factors*: as mentioned above.
- *Signs*: subnormal temperature, cold skin, pale mucosa, wetness around the mouth, collapse of superficial veins, slowness and irregularity of the heart, mild convulsive movements and periodic apnea.
  - Diarrhoea is usually not evident although the abdomen may be slightly distended; auscultation may reveal fluid splashing sounds suggesting a fluid-filled intestine; diarrhoea in which the faeces are profuse and watery to pasty, usually pale yellow to white in colour, and occasionally streaked with blood flecks and very foul-smelling. The course of the disease is often acute and has already been fatal before diarrhoea occurs.

- *Necropsy findings*:
  - Findings: dehydrated carcass, flaccid and fluid-filled intestine, distended and fluid-filled abomasum (which may contain milk clots). Numerous small haemorrhages on abomasal mucosa, hyperemic intestinal mucosa, mesenteric lymph nodes – oedema. Mild atrophy or even fusion of jejunal and ileal villi is often seen, bacilli adherent to the brush borders of enterocytes. Increased epithelial cell loss from the villi. Presence of pseudomembranous ileitis, mucohaemorrhagic colitis and proctitis.
  - Samples for confirmation of diagnosis:
    ▷ Bacteriology: segments of and contents of the ileum and colon.
    ▷ Histology: formalin-fixed duodenum, jejunum, ileum, colon and mesenteric lymph node.
- *Differential diagnosis*, *treatment* and *prophylaxis*: as mentioned above.

Enterohaemorrhagic E. coli

By and large, the disease picture is as named above. In calves kept longer than 10 days on a dairy farm before being sold as *veal calves*, a haemorrhagic enteritis due to *E. coli* may occur. This is most frequently a fatal syndrome.

Possible signs are: anorexia, fever, diarrhoea with mucus containing faeces which become bloody in the later stages, haemorrhagic diathesis on the conjunctivae and mucous membranes of the mouth and nose. Affected calves are depressed, weak, ataxic and sometimes recumbent and appear comatose The best treatment is to give sodium bicarbonate intravenously. Prophylaxis comprises the vaccination of the dam 8 to 4 weeks before calving, and good hygienic conditions.

Necrotoxigenic E. coli

See under enterotoxigenic *E. coli* and enterohaemorrhagic *E. coli*.

10.3.2. Zoonotic aspects of collobacillosis

Cattle have been implicated as a source of *E. coli* which infect and cause food borne disease in humans. Several strains of enterohaemorrhagic *E. coli* associated with enteric disease in humans produce a verotoxin also known as a Shiga-like toxin.

In the last 5 to 10 years, Shiga toxin-producing *Escherichia coli* (STEC), and especially *E. coli* O_{157}, has been shown to be an important zoonotic agent causing haemorrhagic colitis in humans, with potentially further complications, such as the hemolytic uraemic syndrome particularly in children (Karmali *et al.*, 1985; Tesh and O'Brien,

1991). Cattle act as the main reservoir for *E. coli* O_{157} without showing disease signs themselves (Heuvelink *et al.*, 1998). The main routes of infection are considered to be beef products, raw milk and direct contact with farm animals (Heuvelink *et al.*, 2002; Van Duynhoven *et al.*, 2002).

The majority of the *E. coli* O_{157}-isolates obtained from dairy cattle and veal herds had one or two *stx*-genes encoding for Shiga toxin-production and can thus be considered potentially pathogenic to humans. The crude prevalence estimates for *E. coli* O_{157} in dairy herds in e.g. the Netherlands increased in recent years to a prevalence of 14% in 2002. This increase of STEC O_{157} in dairy cattle and veal herds may represent an increased risk of STEC O_{157} infections in humans (Bouwknegt, 2004).

10.4. Treatment of infectious calf diarrhoea

Intensive critical care is required for the treatment of neonatal coliform septicemia. Early identification of septicemia and early therapeutic intervention can improve treatment success. While *E. coli* may be cultured from the blood of septicemic calves, a significant percentage of isolates are gram-positive which justifies the use of antimicrobials which have a broad-spectrum. The ideal protocol comprises the isolation and/or culturing of micro-organisms from blood samples and a sensitivity test for antimicrobials, though this may not always be practical under field conditions. Antimicrobials are given parenterally (intravenously), more than once daily and daily until recovery is apparent. Intravenous fluid and electrolyte therapy are administered continuously until recovery is apparent. Whole blood transfusions could be used in calves especially when immunoglobulin deficiency is suspected from the history or is determined by measurement of serum IgG in blood samples; however, large quantities of transfusions are required.

10.4.1. The general protocol for clinical outbreaks of infectious calf diarrhoea

- The veterinarian should make a visit to the farm.
- He should make an inventory of risk factors.
- He has to examine each risk factor and the way in which that factor can be eliminated or reduced.
- He has to examine the affected animals, as well as the healthy herd mates.
- All affected animals should be identified and isolated if possible.
- All affected animals are to be treated as necessary (see above for details).
- Laboratory samples are to be taken from affected and healthy animals, or calf-side tests are executed.
- Recommendations for the control of diarrhoea in animals to be born in the near future must be made.

- Prepare and submit a report to the owner describing the clinical and laboratory results and how the disease can be prevented in the future (*risk management, see below*).

10.4.2. Considerations regarding the treatment of calf diarrhoea

The veterinarian in the field has to decide whether calves, which are affected with enteric colibacillosis, are bacteremic or septicemic. Commonly, more animals are treated parenterally than strictly required to avoid deaths from septicemia. However, the time does not allow a pre-treatment culture of the causative organisms and determination of the drug sensitivity, so that broad-spectrum antimicrobials are applied. Calf-side tests can – partly – solve this problem.

The advantage of parenterally administered antimicrobials is that the entero-hepatic circulation provides a level of the drug in the intestinal lumen, which may not require the oral administration of the drug.

Multiple antimicrobial drug resistance does occur in *E. coli* and other enterobacteria, when drugs are used on a continuous prophylactic basis.

There is evidence that *E. coli* isolated from the faeces of young farm animals with diarrhoea has developed resistance to trimethoprim-containing antimicrobial products which have been used widely for the treatment of diarrhoea.

Antimicrobials have been used extensively for the treatment of colibacillosis in calves. Some preparations consist of a single antimicrobial, while others are combinations with or without absorbents, astringents and/or electrolytes. It has been difficult to evaluate the efficacy of antimicrobials for the treatment of enteric colibacillosis because of the complex factors affecting the outcome in naturally-occurring cases. These include the presence of concurrent infections, the effects of whether or not milk is withheld from the diarrhoeic calves, the effects of the immune status of individual calves, the variable times after the onset of diarrhoea when the drugs are given, the possible presence of antimicrobial resistance, and the effects of supportive therapy such as fluid and electrolyte therapy.

10.4.3. Oral and intravenous fluid and electrolyte therapy

Oral and intravenous fluid and electrolyte therapy is indicated for:
1. calves in the early stages of diarrhoea;
2. after they have been successfully re-hydrated following parenteral fluid therapy.

- Severely dehydrated or moribund calves may not respond favourably to oral fluid therapy alone, and are too weak to drink or to swallow, or do not drink at all.
- Farmers must be encouraged to provide oral fluid and electrolyte therapy to diarrhoeic calves as soon as possible after the onset of diarrhoea.
- Oral fluid and electrolyte therapy is effective in colibacillosis of calves, because glucose continues to be absorbed by the small intestine by an active transport mechanism accompanied by glucose-coupled sodium absorption and absorption of water.
- In enterotoxigenic (ETEC) colibacillosis, while there is net hyper secretion caused by the enterotoxin, the intestinal mucosa is sufficiently intact: so that water and sodium will be absorbed in the presence of glucose.
- The calf must be returned gradually to a milk diet within a few days in order to avoid the effects of malnutrition.

Extent of dehydration and oral rehydration and electrolyte therapy

Diarrhoea in young calves is still an important disease in young stock. Recently, a survey showed that calf mortality in the USA mounted up to about 9% per year, while about 62% of these were due to preceding diarrhoea (Heinrichs and Kehoe, 2009).

Because young calves may dehydrate very rapidly, it is paramount to closely monitor calves in their first weeks of life. Farmer and employees must be able to detect diarrhoea and dehydration quickly in order to administer a treatment – when needed – as soon as possible. When diarrhoea persists irrespective of oral rehydration, the veterinarian should be called, who will take faecal samples for microbiology laboratory diagnosis, and start a antibiotic treatment if indicated.

Diarrhoea may cause a daily loss of 5 to 10% of body weight in affected calves. Fluid losses over 8% requires an IV rehydration, while losses over 14% may be lethal. A rule of thumb for clinically assessing the extent of dehydration in calves is given in Table 10.3 (after Heinrichs and Kehoe, 2009).

Dairy farms should have a Standard Operating Procedure (SOP) for the diagnosis and treatment of diarrhoea in young calves. This SOP comprises the '*when*' to use oral rehydration and the '*how much to give*' of oral rehydration, as well as other management measures. See Annex 12 for additional information.

Oral electrolyte solutions containing acid phosphate salts are undesirable because they cause net acidification of the blood with a fall in blood pH.

Table 10.3. Clinical signs in aid to assess the extent of dehydration in young calves.

% of dehydration	Clinical signs observed in affected calves
5-6	Diarrhoea, persistent suckling reflex, no other signs
6-8	Slight depression, neck skin fold stands for 2-6 sec, calf is still suckling, sunken eyes, weak calf
8-10	Depressed calf, laying down, very sunken eyes, dry gums, neck skin fold stands for over 6 sec
10-14	Calf does not stand, extremities are cold, neck skin fold does not flatten when tented, comatose
>14	Death

Oral rehydration solutions should not be confused with electrolyte solutions, the latter being more frequently administered to calves *after* weaning and in cattle under stress conditions.

Table 10.4 comprises the advised concentrations of ingredients in oral rehydration products (after Heinrichs and Kehoe, 2009).

In Annex 12 one can find 3 practical scoring tables for assessing the severity of calf diarrhoea and determining treatment under field conditions (Heinrichs and Kehoe, 2009).

Table 10.4. An overview of ingredient concentrations for rehydration solutions (as reference).

Ingredient	Concentration mmol/l	Concentration g/l	Remark
Sodium	70-<145	<3.3	Body water regulator
Sodium bicarbonate	50-80	4.2-6.7	Buffering metabolic acidosis
Sodium citrate	50-80	14.7-23.5	Buffering metabolic acidosis
Sodium acetate	50-80	6.8-10.9	Buffering metabolic acidosis
Potassium	20-30	0.8-1.2	For maintaining pH
Chloride	50-100	1.8-3.5	For maintaining pH
Glucose (dextrose)	<200	<36	As energy source
Glycine	<145	<10.9	Absorption enhancer

Intravenous fluid therapy

Diarrhoeic calves older than 8 days of age could be nearly twice as acidotic as younger calves, and require more bicarbonate to correct the acidosis. It is estimated that *sternally recumbent* calves under 8 days of age require 1 litre of isotonic sodium bicarbonate to correct the acidosis in addition to the necessary quantities of saline to correct the dehydration; for calves older than 8 days of age, 2 litres of isotonic sodium bicarbonate are required.

For *laterally recumbent* diarrhoeic calves under and over 8 days of age, 1 and 3 litres are required, respectively. Isotonic sodium bicarbonate is a safe IV solution, and when uncertainty exists about the degree of acidosis, up to several litres can be used safely for the correction of acidosis and volume depletion.

An equal mixture of isotonic saline (sodium chloride 0.85%), isotonic sodium bicarbonate (1.3%) and isotonic dextrose (5%) is a simple but effective product for parenteral use in severe cases of dehydration and acidosis.

The bicarbonate requirements are calculated using the formula (Radostits, 2001):
$$Weight\ (kg) \times Base\ Deficit\ (mmol/l) \times 0.6(Extracellular\ Fluid\ Space).$$

The base deficit will range from -5 to -20 mmol/l with an average of about -15 mmol/l. The bicarbonate requirements for a 45 kg calf with a base deficit of 15 mmol/l, are $45 \times 15 \times 0.6 = 405$ mmol, requiring 33.75 g of sodium bicarbonate(1 g of sodium bicarbonate yields 12 mmol of bicarbonate). This can be delivered in 2.5 litres of 1.3% isotonic solution.

For moderate dehydration (6 to 8% of BW) fluids should be replaced as follows: hydration therapy 50 ml/kg BW IV in the first 1 to 2 hours at the rate of 50 to 80 ml/kg BW per hour followed by maintenance therapy as described above.

For severe dehydration (10 to 12% of BW) fluids should be replaced as follows: hydration therapy 100 ml/kg BW IV in the first 1 to 2 hours, at the rate of 50 to 80 ml/kg BW per hour, followed by maintenance therapy at 140 ml/kg BW over the next 8 to10 hours at the rate of about 20 ml/kg BW per hour.

For example, a 45 kg calf which is 10% dehydrated should receive 4.5 litres of fluid in the first 1 to 2 hours as hydration therapy followed by 6 to 8 litres of fluid over the next 8 to 10 hours.

Initially, both the acidosis and the dehydration can be treated by the use of isotonic sodium bicarbonate, followed by the use of a combined mixture of isotonic saline and isotonic sodium bicarbonate or multiple electrolyte solutions for maintenance therapy.

Maintenance therapy may be provided using oral rehydration fluids, if the calf is well enough to suck from a nipple bottle or drink from a pail. The use of solutions containing potassium chloride is sometimes recommended on the basis that total potassium stores may be depleted in severely affected calves.

However, they should be used with caution because a severe hyperkalaemia may be present when there is a severe acidosis. If the acidosis and hypoglycaemia are corrected with glucose and bicarbonate, the administration of potassium may be beneficial in restoring total potassium levels. However, solutions containing potassium can be cardiotoxic, particularly if the renal function is not restored. There is no information available on its use in naturally occurring endotoxaemia associated with coliform septicaemia in calves.

Some calves from 10 to 20 days of age with a history of diarrhoea in the previous several days may be affected with metabolic acidosis without obvious clinical evidence of dehydration. They are ataxic, weak, sometimes recumbent and may appear comatose. The IV administration of 2 to 3 litres of isotonic (1.3%) sodium bicarbonate results in recovery within an hour. Hypertonic saline solutions containing 2,400 mOsmol NaCl/l, at a rate of 4 to 5 ml/kg BW IV over a period of 4 minutes has been advocated for the treatment of experimentally-induced endotoxic calves (Radostits, 2001).

10.4.4. Supportive therapies

Additional treatment:

a. *Antimotility drugs and intestinal protectants* There is no strong, scientific evidence that these additional treatments have a beneficial effect.
b. *Alteration of the diet* It is still an issue of debate whether newborn diarrhoeic calves should be given milk or not. In the field it is common practice to limit the milk intake of diarrhoeic dairy calves for 1 or several days, or until there is clinical improvement. Stopping milk supply to diarrhoeic calves is based on the assumption that lactose digestion is impaired and that 'resting' the intestine for a few days minimises additional osmotic diarrhoea caused by fermentation of undigested lactose in the large intestine. In favour of continuing the milk supply is that the intestinal tract requires a constant source of nutrition (e.g. energy) which it receives from the ingesta in the lumen of the intestine. Starving diarrhoeic calves

from milk can result in malnutrition, suboptimal growth rates, and prolonged recovery.

In the field it is also common practice to use oral fluids and electrolytes as milk replacement during the period of diarrhoea. Such fluids are inexpensive, easy to use, readily available and, if used by the farmer when diarrhoea is first noticed, will usually successfully treat existing dehydration and prevent further dehydration and acidosis. Following recovery, calves should be offered reduced quantities of whole milk three times daily (no more than a total daily intake equivalent to 8% of body weight) on the first day and increased to the normal daily supply in the next few days. Milk should not be diluted with water as this may interfere with the clotting in the abomasum.

10.4.5. Prognosis following treatment

One important determinant for the survival of calves affected by enteric colibacillosis is the serum immunoglobulin G (IgG) status of the animal before it develops the disease. The prognosis is unfavourable if the level of immunoglobulins IgG is low at the beginning of the diarrhoea, irrespective of an intensive fluid and antimicrobial therapy being applied or not.

This has stimulated interest in the possible use of purified solutions of bovine gammaglobulin in diarrhoeic calves which are hypogammaglobulinemic. However, they must be given by intravenous route and in large amounts, the cost of which would be prohibitive for its use in practice.

Whole blood transfusion for severely affected calves may be used as a source of gammaglobulins but – unless given in large quantities – would not significantly elevate the serum immunoglobulin levels in deficient calves.

10.5. Prevention of infectious calf diarrhoea

10.5.1. Vaccination of the dam

The immunisation of calves against colibacillosis, Rota- and Corona virus by vaccination of the pregnant dam is a feasible practice in dairy farming (Radostits 2000; 2001; 2007). The oral administration of a K99+-specific monoclonal antibody to calves during the first 12 hours after birth may be an effective method of reducing the incidence of fatal enterotoxigenic colibacillosis particularly when outbreaks of the disease occur in unvaccinated herds.

Vaccines with both the K99-antigen of enterotoxigenic *E. coli* and the Rotavirus, and in some cases the Corona virus, have shown variable results. To be effective the Rotavirus and Corona virus antibodies must be present in the milk after the colostrum period for several days, that is during the period when calves are most susceptible to the viral infection. Vaccination of pregnant cows twice during the dry period at intervals of 4 weeks can increase the colostral antibody levels to *E. coli* K99+. Much lower increases occur at the levels of Corona virus and Rotavirus. Each year thereafter they must be given a single booster vaccination.

The decision to vaccinate will depend on the recognition of risk factors. Such risk factors include those which have been highlighted earlier. Risk management is at least as important as vaccination itself.

Vaccination is an aid to good management and never the replacement for inadequate management.

Chapter 11. Reflections about IBR (BHV-1 virus) infections

11.1. BHV-1 (Bovine Herpes Virus-1; Infectious Bovine Rhinotracheitis, IBR)[2]

Bovine Herpes virus Type 1 (BHV-1) is the viral factor that causes:
1. infectious bovine rhinotracheitis (IBR);
2. infectious pustular vulvovaginitis (IPV);
3. infectious pustular balanoposthitis (IPB).

11.1.1. Main virus characteristics

After infection, BHV-1 virus establishes latency in the neuronal cells of the sensory ganglia that innervate the mucosa at the site of inoculation. The immune system cannot clear the latent virus and reactivation can occur later on in life spontaneously and is usually induced by stressful situations (calving, transport, infections, etc.). Therefore any infected animal should be considered as infected for life and a potential shedder of BHV-1 virus.

11.1.2. The disease

Introduction of the virus in a naive herd results in a very rapid spread of the infection, with morbidity ranging between 20% and 100%, while the mortality level is very variable and may reach up to 15%.

Transmission can take place through direct contact with animals shedding the virus, by aerosol transmission (e.g. in poorly ventilated houses), through semen and embryo-transfer, by contaminated equipment or clothes of e.g. nutritionists, AI technicians, feed suppliers or veterinarians.

An acute IBR infection is characterised by fever, depression, loss of appetite, hyperthermia of the mucosae (*red eye*), mucosal lesions and discharge, dramatic drop in milk production, and in pregnant cows abortion may occur. Infections may occur sub clinically. Opportunistic bovine respiratory disease (BRD) infections may occur simultaneously (see further down).

The disease is still important given the trade-barriers for semen, AI-bulls and other high genetic merit cattle, when infected.

---

[2] Text summarised from and adapted after Franken (2009).

11.1.3. Marker vaccines and diagnostic tests

The current marker vaccines available commercially are based on the deletion of the glycoprotein gE. Vaccination of cattle with these marker vaccines will not induce antibodies against gE while infection with the field virus strains will induce a positive antibody response against the gE. The usefulness of a marker vaccine depends on the existence of a reliable parallel diagnostic test to evaluate the results of e.g. eradication programmes.

There are currently two types of IBR ELISA tests:
1. The first type detects the presence of antibodies against gB, therefore these tests cannot differentiate between field antigen and vaccination antigen with both marker and non-marker vaccines.
2. The second one detects antibodies against gE enabling it to differentiate between field infections and vaccination with a marker vaccine.

Both gB and gE antibodies can be measured in serum and individual or bulk tank milk samples. Antibodies against gB are detectable between 11 and 42 days after infection, while gE antibodies are detected 2-3 weeks later.

Vaccination of cattle with an IBR marker vaccine has a dual purpose: first to reduce the severity of clinical signs due to a BHV-1 infection and second to reduce virus circulation when there is a new introduction of IBR in the herd or reactivation of a latent infection. This second objective is very important in IBR eradication programmes because the final aim of the scheme is to reduce the number of gE positive animals to a minimum. If these animals contact field virus, their gE negative status will change and from that moment they will be considered infected.

11.2. IBR eradication issues

11.2.1. National level

In some countries the prevalence of IBR is very high and culling of IBR positive cattle is economically not feasible. In countries with a low prevalence, such as in the Nordic countries, the strategy of choice for eradication was testing-and-culling.

But in countries with a high proportion of seropositive cattle, the only feasible way to eradicate IBR is to first reduce the prevalence of seropositives in the national herd. The use of *DIVA* (Differentiating field Infected from Vaccinated Individuals) vaccines, also known as marker vaccines, can be a useful tool in this process. Animals are vaccinated twice yearly (spring and autumn).

The speed of the eradication programme will depend mainly on the initial herd sero-prevalence, the culling rate and the rate of new infections/re-activations.

11.2.2. Herd level

Control strategies:

For controlling IBR, there does not exist a universal strategy to follow in each and every country. We present a few strategies as an example:
1. In some countries free from IBR, vaccination is not allowed. Bulk tank analysis, and individual blood testing will give the status of the farm. ELISA tests must be done and positive animals will be culled immediately.
2. In other countries, where the farmer does not use marker vaccines but only conventional vaccines, it will be impossible to differentiate from vaccinated animals and naturally infected animals. The antibodies remains for more than 5 years.
3. If the farmer does not vaccinate at all, it is clear that using ELISA gB diagnostic tests facilitates the detection of positive animals and negative animals. If there are up to 10% of positive animals, eradication is feasible by culling the positive animals.
4. If marker vaccines are authorised, the ELISA gE negative test could be used and negative animals will be protected. If we like to achieve an IBR free status after some years (normally 5 years after starting the programme it is expected to have less than 5% of positive animals) we can cull those animals from the herd. But this depends heavily on whether or not we have a closed farm (biosecurity measures) and whether or not it is sufficiently isolated from other farms.
5. Bulk tank analysis and blood tests must be done to set the diagnosis. Biosecurity plans must be in place. All young animals must be tested, and if the farmer buys new animals, it is not enough to conduct only one single test. It can be negative, because the virus can remain in the lymph nodes or in some nerves. Due to this phenomenon, the farmer should buy his new animals from IBR free herds. The application of glucocorticoids in such animals will induce clinical disease, as stress will.

Vaccination strategies at herd level:

The strategy is to vaccinate – if allowed – with gE deleted vaccines (alive or attenuated). This will make the control more easy. But this is not the case for all farms and countries. The gB vaccines are used in several countries where no gE deleted vaccine is allowed.

One of the principles is to vaccinate all animals twice a year, and starting as soon as possible.

Management of IBR outbreaks (again, there is no panacea available):

- Case 1- Free countries. Application of test and cull or eradication when it is not allowed to vaccinate first.
- Case 2- In countries where marker vaccines are allowed: start vaccination with marker vaccines. Live vaccines must be used intra-nasally in acute outbreaks, for interferon stimulation.
- Case 3- Countries where only gB vaccines are available. Emergency vaccination and antibiotics therapy to prevent secondary infections – like *M. haemolytica, P. multocida,* or *H. somni.* There remains the risk of other virus infections.

Remark: it is good veterinary practice to sample young stock to see whether they are positive or negative. This gives also a good feedback about the vaccination efficacy. Note that prior to 6 months maternally derived antibodies may hamper the proper diagnosis.

An IBR free herd surrounded by infected herds is at high-risk and in order to maintain a good immune status in these animals, it is important to prevent a severe IBR outbreak should a new infection take place, e.g. good biosecurity and vaccination.

Monitoring of an IBR free status varies between countries. Bulk milk testing or individual blood sampling is carried out every 6 to 12 months depending on the country (see below).

Control: once a herd achieves an IBR free status, farmers will aim to avoid new infections entering the farm by buying cattle exclusively from other, IBR-free certified farms and applying additional biosecurity measures as well as good hygiene control.

Long distance transport, a new environment with a different climate, mixing of stock, a change of diet and a different treatment create severe stress. This means the animals are more susceptible to diseases like IBR. Cattle that are vaccinated with the IBR marker vaccine will be protected against IBR.

In addition, all *biosecurity plans* must be operational (see Section I in this book).

Make sure that blood samples are taken at the farm or farms or origin, from which the animals originate to make sure that they are BHV-1 gE-free. Upon arrival, cattle must be put into quarantine for at least 3 weeks, and sampled for BHV-1 gE testing.

Remark: note that maternally derived antibodies (MDA) against BHV-1 virus will hamper the proper detection of an actual IBR infection in calves when these are sampled before the age of 6 months.

11.2.3. Marker vaccination

A cow infected with the IBR-virus forms antibodies against the virus. With laboratory tests these antibodies can be detected in milk and blood.

Vaccination of a cow with the conventional vaccine, results in the production of antibodies that are similar to antibodies against the field virus. Therefore, milk and blood tests do not discriminate between infected and conventionally vaccinated animals. The IBR marker vaccine (Figure 11.1) however, does not contain the glycoprotein E (gE-deleted vaccine) which is part of the IBR virus. If cows are tested for IBR with the gE-ELISA test, it is possible to detect infected animals, while vaccinated non-infected animals will be identified as IBR-free. The test can be run in serum and milk.

Figure 11.1. The marker vaccine principle (after Franken, 2009a,b).

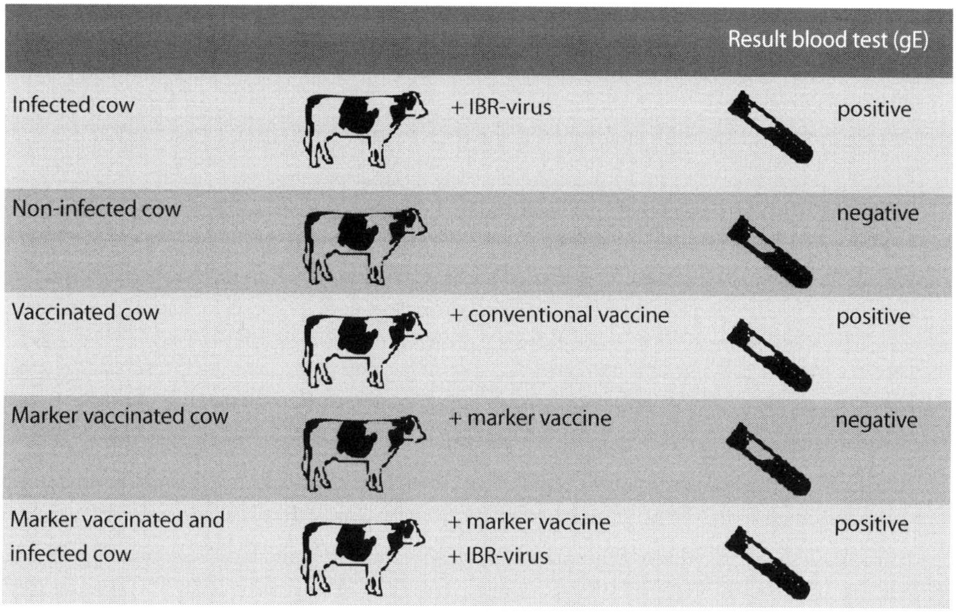

			Result blood test (gE)
Infected cow		+ IBR-virus	positive
Non-infected cow			negative
Vaccinated cow		+ conventional vaccine	positive
Marker vaccinated cow		+ marker vaccine	negative
Marker vaccinated and infected cow		+ marker vaccine + IBR-virus	positive

11.2.4. Epidemiological issues regarding the certification of IBR-free herds

Not all infected animals, nor all infected herds can be detected due to the diagnostic test characteristics which are never 100%; however, newly infected herds can be detected in time when one executes regular monitoring of the herds through bulk milk tank sampling. Note that maternally derived antibodies against BHV-1 virus hamper the proper detection of an actual IBR infection in calves when these are sampled before the age of 6 months.

Follow the O.I.E. guidelines and manual (see OIE website).

On a national and regional basis it would be (politically, socially and economically) highly valuable to know what the risk is at a given moment in time of a (more or less severe) outbreak, and whether detection and control measures would be adequate to avoid spread to other farms.

Not all countries have procedures that are identical to those in the O.I.E. guidelines. Regardless of what method of qualification or monitoring is chosen, it should specify scientific data about:
- the risk of an outbreak;
- the detection of an outbreak before transmission to other herds takes place.

There must be a balance between scientific justification, objectives, methods, practicality on the one hand, and politics, socio-economics, costs on the other hand, for example when estimating a necessary sample size for testing.

The standards for monitoring IBR-free herds vary per country, for example:
- Country A: Testing blood samples twice a year
- Country B: Testing bulk milk samples every 6 months
- Country C: Testing bulk milk samples monthly
- Country D: Testing bulk milk varies per area

11.2.5. Potential risks of outbreaks of IBR

The sensitivity of a diagnostic test (how effective the test is in identifying animals with the disease) is not 100 %. So, herds with positive animals that test false-negative may get a certificate. This means that an outbreak in a certified herd can occur, which can lead to outbreaks in other certified herds if not detected in time. The frequency that this might occur is unknown.

How to reduce these risks?

- intensify the intake procedure (testing 2 or more times at a certain interval);
- increase sensitivity of the intake test by e.g. serial or parallel testing;
- intensify the herd and animal monitoring procedures;
- put further restrictions on contacts (test animal before it goes to another certified herd);
- institute vaccination programmes.

11.2.6. IBR – aspects of epidemiological modelling

The IBR epidemiology regards the within herds dynamics of infection, as well as the between herds dynamics of infection, because it is determined by: animals, herds, animal movements (EU directive 64/432), other contacts, IBR virus features (incubation and reactivation of the herpes virus), detection, vaccination.

Epidemiology modelling research shows that – without vaccination – one infected cow/herd will transmit the infection to (on average) 5.6 other cows/herd (= R-value). This means that once one animal gets infected, the virus spreads quickly through the whole herd.

Vaccination and other measures that reduce the risk of an IBR infection will decrease this R-value. In an effective eradication programme, one outbreak in an IBR-free herd will (on average) induce less than one subsequent outbreak, which means that the R0-value between herds must be below one. Qualifying and monitoring of IBR-free herds must be justified in this respect.

BHV-I-infected cattle can reactivate BHV-I virus especially under stress conditions. In some cases when transmission occurs, the virus can spread in the herd. In a herd this risk depends on the number of infected animals, on whether they are tested sero-positive or false negative, on the reactivation rate, the herd size and the vaccination status, general immune status. Based on these data a risk can be calculated for closed herds.

Table 11.1 indicates that in case of a single or a few infected animals only a small risk of an outbreak exists (1-5%). In the qualifying of IBR free herds the whole herd testing does eliminate largely this small effect of the sero-positives. Because of the imperfect tests there is always the risk of a false negative test result (Franken, 2009a,b).

Table 11.1 also shows that in case of a herd of 100 cattle with a prevalence of 30%, the risk of circulation due to reactivation increases drastically up to 25%. In even larger

Table 11.1. Risk of an outbreak of IBR beacuse of reactivation (closed herd) and size of the outbreak.

Vaccination	Number of seropositive cattle	Herd size (number of cattle (n))										
		40			60			80			100		
		no	dead	live	no	dead	live	no	dead	live	no	dead	live
	0	0%	0%	0%	0%	0%	0%	0%	0%	0%	0%	0%	0%
	1	1%[a]	1%[a]	0%[b]	1%[a]	1%[a]	0%[b]	1%[a]	1%[a]	0%[b]	1%[a]	1%[a]	0%[b]
	2	3%[a]	1%[a]	0%[b]	3%[a]	1%[a]	0%[b]	3%[a]	1%[a]	0%[b]	3%[a]	1%[a]	0%[b]
	3	4%[a]	2%[a]	0%[c]	4%[a]	2%[a]	0%[b]	4%[a]	2%[a]	1%[b]	4%[a]	2%[a]	1%[b]
	4	5%[a]	3%[a]	0%[c]	5%[a]	3%[a]	1%[b]	5%[a]	3%[a]	1%[b]	5%[a]	3%[a]	1%[b]
	5	6%[a]	3%[a]	0%[c]	6%[a]	3%[a]	1%[c]	6%[a]	3%[a]	1%[b]	6%[a]	4%[a]	1%[b]
	10	10%[a]	5%[b]	0%[c]	11%[a]	6%[a]	0%[c]	12%[a]	6%[a]	1%[c]	12%[a]	7%[a]	1%[c]
	15	12%[a]	5%[b]	0%[c]	15%[a]	7%[b]	0%[c]	16%[a]	8%[b]	0%[c]	16%[a]	9%[a]	0%[c]
	20	11%[b]	3%[b]	0%[c]	17%[a]	8%[b]	0%[c]	19%[a]	10%[b]	0%[c]	20%[a]	11%[b]	0%[c]
	25	5%[b]	0%[c]	0%[c]	17%[b]	7%[b]	0%[c]	21%[a]	10%[b]	0%[c]	23%[a]	12%[b]	0%[c]
	30	0%[c]	0%[c]	0%[c]	16%[b]	5%[b]	0%[c]	22%[a]	10%[b]	0%[c]	25%[a]	13%[b]	0%[c]
	40				3%[c]	0%[c]	0%[c]	20%[b]	6%[b]	0%[c]	27%[b]	12%[b]	0%[c]
	50				0%[c]	0%[c]	0%[c]	10%[b]	0%[c]	0%[c]	25%[b]	8%[b]	0%[c]
	75							0%[c]	0%[c]	0%[c]	0%[c]	0%[c]	0%[c]
	100												

[a] >80% of seronegative cattle will be infected.
[b] 20-80% of seronegative cattle will be infected.
[c] <20% of seronegative cattle will be infected.
(ASG, 2000)

herds with higher proportion of infected animals this risk is up to 75%. Another important outcome of the model calculations as shown in the table is the effect of the live vaccine compared to inactivated vaccine on the risk of an outbreak in larger and high prevalence herds. The explanation is only the difference in Ro: live 1.5; inactivated 2.6. A higher Ro value stands for a better protection of the susceptible part of the population. Although the inactivated vaccine has proven to reduce the reactivation in infected animals, the net effect on the risk of an outbreak is less as compared to live vaccine. In order to get the advantages from both vaccines to improve the results of the live vaccine one should know the infection status of the individual animals in the herd, for instance by serological testing.

Chapter 12. Reflections about BVD (BVD virus) infections[3]

12.1. Introduction

Bovine virus diarrhoea (BVD), can be a highly virulent and contagious disease that causes high morbidity and mortality. There are 2 biotypes of the ruminant BVD virus: cytopathic and *non-cytopathic* based on cellular culture characteristics. The non-cytopathic type is the most common and most important. This type is able to cross the placenta and might cause a persistent infection in the fetus.

12.2. Pathogenesis

The pathogenesis of the BVD virus is dependent on several host factors:
1. immunocompetence or immunotolerance to the virus;
2. age of the animal;
3. transplacental infection and gestational age of the fetus;
4. (passively or actively derived) immune status;
5. genetic different isolates of the virus;
6. presence of stressors (e.g. other infectious cattle diseases, suboptimal feeding).

Usually, most cattle will successfully control a natural infection, develop immunity and eliminate the virus so that latency and shedding does not occur.

12.2.1. Different disease appearances and their most important signs

- *Subclinical BVDV in immunocompetent cattle*: fever, depression, mild diarrhoea, inappetence for a few days. *Source:* persistently infected animals (PI) in the herd.
- *Peracute BVDV in cattle of all ages*: can provoke severe diarrhoea, fever, agalactia and rapid death. Thrombocytopenia and haemorrhagic syndrome in adult cattle and veal calves. *Source:* introduction of (PI) animals into a inadequate vaccinated herd.
- *Reproductive failure*: abortion, still birth, weak neonates, congenital defects, decreased conception rates.
- *Mucosal disease in PI animals*: signs of infection with a cytopathic strain: fever, diarrhoea, oral erosions and high case-fatality rate. PI animals often have a retarded growth and a weak immune system.
- The BVD virus may also cause general immuno-suppression.

---

[3] Text summarised from and adapted after Radostitis (2001) and Cannas da Silva (2002)

12.3. Reproductive failure during different reproductive stages

In general, seronegative virus-free cattle (often heifers) are at risk of a BVD infection that leads to reproductive failure and transplacental infections. Examples are given here:

- *Prior to insemination*: failure of or delayed ovulation.
- *Insemination with BVD infected semen*: poor conception rates; often occurs only at the first insemination of seronegative virus-free heifers.
- *0-40 days gestation*: decreased conception rate and pregnancy rate; often seronegative -heifers.
- *40-125 days gestation*: BVD virus crosses the placenta causing death of the fetus, abortion, mummification, congenital defects or persistent infected calves (PI) with a lifelong infection without signs.
- *125-175 days gestation*: transplacental infections in this gestation period lead to cerebellar hypoplasia and ocular abnormalities (retinal atrophy, optic neuritis, cataract and microphthalmia). Calves born may be smaller and have a curly hair coat.
- *180 days gestation - birth*: approximately after day 150 of gestation a BVD infection induces a immune response and the fetus is able to eliminate the virus. If abortion does not occur, born calves are BVD virus free and are BVD antibody positive.

12.4. Economic impact of BVD

Considerable economic losses are caused by BVD: increased calving interval; infertility; embryonic death; mummification; abortions; calves with malformations; decreased milk production; decreased immune status e.g. more susceptible for other infectious diseases, and persistently infected animals.

To estimate the economic losses of BVD infections in a herd, several factors are important to consider:

- the structure and size of the herd;
- the proportion of susceptible animals in the herd;
- the manifestation (or absence) of clinical signs;
- level of production loss;
- the yearly risk of (re-)infection (e.g. number of PI animals on the farm);
- costs of vaccination and eradication.

The final sum is calculated by adding up the products of the different factors. In practice, all these factors, except the structure and size of the herd, can only be estimated. The percentage of susceptible animals can be estimated by sampling the herd (bulk milk or serology). The risk of infection of these susceptible animals is determined by the prevalence of PI animals and management factors (separate housing of age groups, etc.).

12.4.1. An estimation of losses due to BVD

The level of losses was estimated for an average susceptible dairy herd in Germany (Wolf, 1997). In this average herd, there were 100 animals, 15 calves of 0-3 months, 20 young stock of 4-18 months, 15 heifers and 50 cows. Calving took place throughout the year. An infection is introduced with the birth or the purchase of a PI animal. A calculation of the economic loss from BVD requires modeling the disease in the herd and weighting the different production parameters affected. The spectrum of disease ranges from infertility (return to oestrus, resorption of the fetus, abortion, birth of malformed or weak calves), to respiratory and gastro-intestinal diseases of varying severity. Beside this, different disease signs occur in different age groups on the farm. Also, the morbidity of the different diseases (the proportion of the herd affected) depends on the immune status of the herd. To model the effects of BVDV infection in the herd in the first place, the percentages of different age groups are shown in the column A of Table 12.1.

The susceptibility to BVD of the animals in the different age groups is also given. While the herd is completely susceptible (antibody free), the risk of infection after introduction of a PI animal is 100% (column C).

After the disappearance of colostral antibodies, PI animals will shed more than 100,000 animal infective doses per gram of excreta. Not only direct, but also indirect contact will ensure virus spread. Although transiently infected animals shed less virus than PI animals (e.g. one animal infective dose per gram of excreta over several weeks), the number of these animals is much larger than the number of PI animals. Even in a completely susceptible herd, not all animals will suffer from clinical disease. Post-natal infections often follow a subclinical course. Severity of symptoms, morbidity and mortality also differ with the virulence of different virus strains. It is clear that the greatest damage is caused by abortion, resulting in milk loss (Cannas da Silva, 2002b).

12.5. Persistently Infected (PI) animals

- *Occurrence*: if pregnant seronegative cows get infected with the BVD virus in the period 40-125 days of gestation, one of the possible outcomes is that fetuses get immunotolerant for the BVD-virus, since in that period the immune system of the fetus starts to develop. The virus is not eliminated from the fetus and the calf is born as a persistently infected (PI) calf. Most of the PI animals die within a few months suffering from a weak immune system and rarely live for more than 1-2

Table 12.1. Modelling the epidemiological and economic effects of BVDV infection in a dairy herd.

A Age group	B Proportion of the herd (%)	C Susceptibility (%)	D Clinical signs	E Probability of occurrence of clinical signs (%)	F Costs per case (€)	G Losses per animal (€) (BxCxExF)
Calves 0-3 months	15	100	RDS	25	12.50	0.48
Young stock 4-18 months	20	100	RDS	25	12.50	0.64
Heifers	15	100				
Cows	50	100				
1 month pregnant	5.4	100	Repeat breeding	80	10	0.44
2-3 months pregnant	11	100	PI animals >> mortality	100	153	16.88
4-7 months pregnant	22	100	Abortion	50	307	33.75
			Stunted growth of calves	25	103	5.63
8-9 months pregnant	11	100	Calf mortality/ stunted growth	25	77	2.11
Total average loss per affected animal						59.92

(after Wolf, 1997)

years. But a PI animal may also still give birth to a new PI calf. A PI animal over 2 years of age causes severe problems in the herd because it constantly infects pregnant herd mates.
- *Signs*:
  1. Birth and growth are apparently normal (these animals are rarely detected). They develop the typical form of 'Mucosal Disease', if in contact with heterologous strains of cytopathologic strains (= *super infection*).
  2. Born weak, poorly developed, dying a short while after birth *(differential diagnosis with neonatal asphyxia)*. Congenital lesions may be present: cerebral hypoplasia; cataracts; optic nerve degeneration; brachygnatia; alopecia of various degrees. After birth, they show signs of diarrhoea and/or respiratory distress, or death. Death is usually attributed to other causes than BVD (e.g. *E. coli*, rota-/corona virus, BRSV, *Pasteurella*). Usually, because blood and tissue tests to detect the virus are not made, and hence, the real cause of death is not correctly identified.
  3. The calves can show repeated or chronic disease (kerato-conjutivitis, enteritis, pneumonia, ecto- and endoparasitosis), that does not respond to treatment. The animals show signs of other disease, so BVD remains hidden, e.g. BLAD (Bovine Leukocyte Adhesin Deficiency), chronic injuries, healing failure.
- *Differential diagnosis*: it is always justified to suspect BVD when animals do not react to treatment and or suffer from severe anorexia. Some animals may have prolonged fever, haemorrhages, blood in faeces, diarrhoea and thrombocytopenia. Whenever there are too many abortions, abnormal fetuses, decreased conception rate, BVD should be suspected too. Whenever there are signs of *Mannheimia haemolytica*, *Pasteurella multocida*, *Histophilus somni*, PI-3, bovine respiratory syncytial virus, IBR, viral respiratory infections, or infectious enteritis, a PI calf may be present.
- *Diagnosis*: detecting the virus (ELISA blood samples) in animals that have chronic diseases (e.g. chronic diarrhoea, parasitism) is paramount.

12.6. Diagnosis of BVD – Mucosal disease

- *Occurrence*: when PI calves get infected by a cytopathic strain during the (early) rearing period, they often will develop mucosal disease. Calves born from PI-dams are always seronegative (antibody negative). Presence of PI animals in the herd could be discovered when some animals develop mucosal disease after vaccination with a modified (cytopathic) live vaccine. However, not all PI animals in contact with a cytopathologic strain will develop this clinical picture, because they may have antibodies against some heterologous cytopathologic strains.
- *Signs*: high hyperthermia (bi-phasic fever 40-41 °C), weight loss, poor growth, poor body condition, prolonged and profuse diarrhoea, leucopenia, mucosa

lesions like erosions (mouth, teats, vagina, interdigital space, tongue, oesophagus, abomasums, scrotum), mucopurulent nasal discharge, corneal edema and often death occurs within several days. Sometimes the only clinical sign in a PI calf is limited to swollen mouth papillae, without the animals showing typical erosions of the disease.

- *Differential diagnosis*: confusion may occur with BRD (bovine respiratory disease complex, see elsewhere in this Section II). Diagnosis is 'relatively easy' in calves which show the typical signs of mucosal disease.
- Postmortem *findings*: the typical lesions particularly in the oesophagus (erosions) and also in the gastro-intestinal organs (*mucosal disease*).

12.7. Diagnostic testing and herd screening on BVD

It is important to obtain or exclude the diagnosis of BVD infections in case of a suspected acute infection or the screening of herds (regularly or in case e.g. a PI calf has been detected).

- *Acute infections*: as early as 3 days to 8-10 days after infection. ELISA/PCR antigen screening can be done in serum blood samples. Nasal swabs can be submitted for screening for virus. In herd outbreaks, paired blood samples of 'normal' animals could be collected 30 days apart, in order to detect a four-fold increase in serum antibody titers (seroconversion).
- *Abortion*: in dams without BVD-antibodies, BVD could be ruled out as a cause of abortion. Some aborted calves are antibody positive.
- *Calves with congenital defects*: blood for antibody screening must be drawn before feeding the first colostrum.
- *Persistently Infected calves*: in practice, calves can be tested at dehorning (around 4-8 weeks of age) for antigen presence; with nearly 100% confidence the test-positive calves are PI calves, and hence re-testing is not considered. The RT-PCR allows detection of PI animals and it is not needed to repeat the test.
- *Herd screening*: blood samples from all animals of the herd are tested for the presence of BVD antigen. Calves born during the coming 9 months should also be tested to detect any additional PI animals that are infected intra-uterine at the time the herd infection occurred.
- *Quick scan*: in order to predict whether BVD plays a role on a specific farm, several young animals in the age of 8-12 months are screened for BVD antibodies. Simultaneously the bulk tank milk is tested for BVD antigen by PCR (GD, the Netherlands).

12.8. Considerations regarding the control of BVD at the herd level

A BVD control and elimination programme has 5 phases:
1. *Prevention of introduction of infection onto the herd* Testing all purchased animals for BVD virus antigen, before introducing them into the herd. Furthermore strict biosecurity measures should be carried out.
2. *Identification and elimination of PI animals* Blood samples from all animals of the herd are tested for the presence of BVD antigen. Calves born during the coming 9-12 months should also be tested to detect any additional PI animals that are infected intra-uterine at the time the herd infection occurred. All identified PI calves should be culled as soon as possible.
3. *Vaccination before breeding* The chosen vaccine type and the vaccination programme make the difference in achieving a BVD-free status and is part of any eradication programme. If one is working with killed vaccines – one needs to revaccinate 30 days later. After this primary vaccination, all animals should be vaccinated every 6 months or yearly depending on the vaccine used, to achieve a high antibody level. This is particularly important for heifers before the 1st AI; the dams are fully protected 4 weeks before mating or AI (depending on the vaccine).
4. *Control of BVD* Vaccination is to be followed by control. The most effective method to control BVD is to test all animals for antigen presence. For example, testing the newborn calves is a good method that allows assessment of vaccine efficacy, test-positive animals are for almost 100% certainty identified PI-calves and should be slaughtered as soon as possible.
5. *Screening of free herds* In some countries free from BVD, bulk milk tank screening or regularly random antibody screening in older calves are good control procedures. However, in other countries slaughtering the PI and not vaccinating is a high risk strategy, because full biosecurity measures may not in place and it is common to buy animals from herds or countries of unknown BVD status.

Chapter 13. Summarising other diseases in young stock[4]

(in alphabetical order)

A quick reference guide to normal diagnostic criteria is useful in daily practice. It is presented in Table 13.1.

Table 13.1. Normal diagnostic parameters in young stock.

Diagnostic parameter	Normal range in young stock
Rectal temperature (°C)	38.5-39.5
Respiratory rate (bpm)	20-40
Heart rate (bpm)	70-90
Body condition scores	3-3.5
Conformation and development in relation to age	
Growth performance per age category	See Section I and Annexes
Hair coat appearance	
Turgor (skin fold in the neck)	
Rumen fill scores	3-4
Locomotion scores and posture of back legs	1-2

13.1. Blue Tongue (BT)

- *Occurrence*: infectious disease. Transmitted by *Culicoides* spp. (*imicola, variipennis, fulvis, obsoletus, mudipalpis, orientalis, wadai, actoni, brevitarsis, peregrynus, oxystoma, brevipalpis, fuvus, sonorensis).* Present all over the world, including parts of Europe. Serotype 8 and 4 affecting cows. Serotype 8 much more severe for cattle. Epidemic areas connected with the conditions of the vector spreading. Culicoides can survive in winter periods. Several serotypes according to the area (1-24). Cattle are the reservoir and amplifying host.
- *Causal agents*: *Orbivirus*. Family *Reoviridae*, 24 serotypes worldwide.
- *Signs*: typical blue tongue, loss of appetite, catarrhal stomatitis, rhinitis, enteritis, lameness due to coronitis and myositis, fever, salivation, fetive breath, also found in semen.
- *Diagnosis*: clinical signs. Differential diagnosis with BVD-MD, FMD, MCF, Rinderpest, Rift valley fever, vesicular stomatitis, IBR (Bexiga *et al.* in OIE, 2009).
- Laboratory: Virus-isolation; ELISA; PCR; RT-PCR; RT-qPCR. The two latter are preferred (e.g. for mass detection) (De Clerq *et al.* in OIE, 2009).

---

[4] Text adapted after and from Andrews *et al.*, 1992; Radostits, 2000, 2007; Smith, 2009

- *Treatment and prevention*: symptomatic treatment is indicated and most bovines recover from the disease, depending on serotype. Vaccination – against one or more serotypes – is a very effective method of prevention (combination serotypes 1 and 8 available). Insect repellents like Permethrim; insecticides. Housing animals for the night.
- For a recent update, please check OIE (2009).

13.2. Bloat

- *Occurrence*: chronic tympany occurs in unweaned or recently weaned calves. This may be associated with poor transitioning due to inadequate ruminal development and function. Though hairballs have been found in affected calves their aetiological significance is debatable.
- *Signs*: swelling of the left paralumbar fossa or entire upper flank.
- *Diagnosis*: clinical signs.
- *Treatment*: in acutely affected cases passage of a stomach tube will relieve the tympany. In chronic cases insertion of a permanent ruminal cannula may be required.
- *Prophylaxis*: alter weaning management with earlier introduction of hay and concentrate ration and delay weaning until calves are consuming adequate quantities of these feeds.

13.3. Bovine Respiratory Disease Syndrome, BRD

Differential diagnosis include:

Bovine Respiratory Syncytial Virus, BRSV
- Frequently, the infection is transmitted from older to younger animals.
- Respiratory problems (pneumonia), sometimes sudden death.
- Treatment is intensive (antibiotics; acetylsalicylic acid) and can be expensive. Vaccination is a preventive option.

Histhophilus somni
- Occurs at an age of 1 to 6 months.
- Signs are pneumonia-like, but difficult to differentiate because they form part of a complex of diseases and often refractive to treatment.

Mycoplasma spp. (*haemolytica*)
- May cause a pneumonia in calves.
- Signs are severe pneumonia, inspiratory and expiratory dyspnoea, orthopneic position, sudden death.

- Diagnosis by tracheal lavage swabs and culturing.
- Treatment with broad spectrum antibiotics.
- Vaccination is a preventive measure.

Para-Influenza 3 virus
- Occurs frequently together with BRSV. Vaccination is an option. Treat secondary complications with antimicrobials.

Pasteurella multocida
- Commonly a secondary infection after viruses, causing a severe pneumonia. Treatment with the proper antibiotics. Prevention through vaccination is an option.

13.4. Fatal aplastic haemorrhagic syndrome of calves (bleeding calf syndrome): a new, emerging disorder

Throughout 2008 and 2009, various cases of a so far unknown disease in young calves have been reported in Europe (e.g. Germany, Italy, UK, the Netherlands, Belgium). It regards a *haemorrhagic diathesis* or *bleeding calf syndrome* (e.g. K. Doll – Germany; W. Klee – Germany; D. Black – UK; A. Gentile – Italy; J. Verhoeff – the Netherlands). These calves are commonly born healthy, but develop this syndrome after colostrum intake (changing blood composition: thrombocytopenia; bone marrow damage; pyrexia; spontaneous bleeding) (Fridrich *et al.*, 2009). Calves bleed to death through body orifices or skin lesions. It is thought that the syndrome is not infectious, and it has been hypothesised that the dams develop antibodies to immune cells of their calf. Other hypotheses point to a potential association with BVD or BTV-8 vaccination campaigns, but this is contradicted by the fact that in traditional beef cattle herds in Scotland the syndrome also occurs while no BVD vaccination campaigns have been conducted. So far, no unanimous scientific evidence about etiology and pathology is available (yet). The hypothesis that a circovirus, with high similarities to porcine circovirus type 2b (PCV2b), may be involved is currently being investigated (Kappe *et al.*, 2009). The disorder has recently been named 'Fatal aplastic haemorrhagic syndrome in calves'.

13.5. Infectious Bovine Keratoconjunctivitis (pink eye)

- *Occurrence*: most common and costly ocular disease of cattle; ubiquitous; most common in summer and autumn; could have epizootic cases; transmission: flies and dust. More common in white-faced breeds, e.g. some Hereford and its crossbreeds.

- *Causal agent*: *Moraxella bovis*. Solar radiation, flies and dust are contributing factors.
- *Signs*: conjunctivitis, lacrimation, blepharospasm, photophobia, central corneal opacity.
- *Diagnosis*: culture; pink eye or fluorescent antibodies.
- *Signs*: corneal opacity with small elevation. Lacrimation.
- *Treatment and prevention*: self limiting disease; topical antibiotics; subconjunctival penicillin; parenteral oxytetracyclines. Vaccination.

13.6. Navel and joint illness

These are common problems in the young calf. At birth there is a sudden change from the fetal circulation to that of the newborn calf. The umbilical vessels in the umbilical cord rapidly loose most of their blood but still remain patent, allowing the introduction of infection.

Infection can be caused by a single or a mixture of organisms (see Table 13.2), usually provoked by poor hygienic conditions, inappropriate colostrum management and or an open urachus.

13.7. Parasitic diseases

Strongylus spp.: *Strongylus papulosus*
- *Signs*: dirty stables, emaciation, cachexia. May also occur after colostrum feeding.
- *Diagnosis*: faeces analysis.
- *Treatment*: anthelmintics.

Ascaris: *Toxocara vitolorum*
- Larvae are eliminated in the dam with the first colostrum, and in the pastured or fresh grass fed calf. A galactogenic or true milk infection.
- *Signs*: the typical smell of the faeces, diarrhoea of the calf, sometimes colic. Is a sporadic infection.
- *Diagnosis*: faeces analysis (parasites present in faeces).
- *Treatment*: anthelmintics.

Coccidiosis: *Eimeria zuerni*, *Eimeria elipsoidalis*, *Eimeria bovis*
- Infection by oocysts ingestion.
- *Signs*: hypothermia or normal temperature, anaemic animals, faeces with blood, decolouration of mucosae, ataxia, recumbence, death.
- *Diagnosis*: through microscopic faeces examination.
- *Treatment*: coccidiostatics, not in calves weighing over 80 kg/ after weaning.

Table 13.2. Overview of some micro-organisms predominantly involved in navel illness.

Streptococcus spp
- *Signs*: local inflammation, swollen navel, pain, localised peritonitis, depression, pyrexia, tachypnoea, varying degree of dehydration, acidosis, recumbence and death; inappetitence, dullness, urachus infection- animal unthrifty and rather slow to move, endocarditis, panophthalmitis with hypopyon, meningitis and nystagmus, hyperaesthesia and tonic-clonic convulsions
- *Diaginosis*: swollen navel, differential diagnosis with hernia umbilicalis

Escherichia coli
- *Signs*: diahrroea, sudden death, swollen articulations
- *Diagnosis*: swollen navel, differential diagnosis with hernia umbillicalis

Erysipelothrix insidiosa
- *Signs*: depression, pyrexia, hyperthermia
- *Diagnosis*: swollen navel, differential diagnosis with hernia umbillicalis

Pasteurella multocida
- *Signs*: together with pneumonia
- *Diagnosis*: swollen navel, differential diagnosis with hernia umbillicalis

Actynomyces pyogenes
- *Signs*: swollen navel, abcess
- *Diagnosis*: swollen navel, differential diagnosis with hernia umbillicalis

Fusobacterium necrophorum
- *Signs*: swollen navel, abcess, necrotic tissue
- *Diagnosis*: swollen navel, differential diagnosis with hernia umbillicalis

Adapted after Andrews *et al.*, 1992; Radostits, 2000, 2007; Smith, 2009

Dictyocaulus viviparous and other spp. (lungworm)
- *Occurrence*: in pastured or fresh grass-fed calves and maiden heifers in pastures with a history of lungworm. Often in pastures used exclusively every year for rearing young stock. Age immunity occurs in exposed animals but not if unexposed hence the condition may occur in older young stock also.
- *Signs*: coughing, dyspnoea, illthrift, loss of body condition.
- *Diagnosis*: clinical signs, age of animals, husbandry practices, larvae in faeces, eosinophilia in blood sample.
- *Treatment*: anthelmintics some with residual activity (Note: never use these in heifers just prior to lactation).
- *Prophylaxis* (applicable to all parasite infections above): pasture management to avoid build up of parasite, early prophylactic repeated anthelmintic administration, vaccination.

13.8. Paratuberculosis (Johne's disease)[5]

Agent:
- *Mycobacterium avium* subspecies paratuberculosis, also known as Map. It is very common world wide and affects usually *adult animals*. In one study, the most common animals to be detected as Map-positive using individual animal faeces culture were older Holstein-Friesian cows and beef bulls, particularly Limousins (Richardson *et al.*, 2009). The culture of faeces is the best method for detection of the disease in a farm. Antibody ELISA kits are also available but they can give false-negative results (low sensitivity). It means that a test-negative animal today can be test-positive 1 month later and after another month again test-negative.
- The incubation period is long; most frequently calves are infected after calving, and they will be test-positive after 12 to 18 months, and then can show clinical signs.

Clinical findings:
- Silent infection in calves, heifers and young stock up 2 years of age. No clinical signs and no effects on growth performance. No influence on body condition scores. They can shed the agent. Possible detection in faeces. Subclinical disease occurs in adults as carriers; 15 to 25% can be positive to individual faeces culture. Normally negative in all serological tests.

Clinical disease:
- Clinical signs are typically detected at 4 years of age (SD 1.6 yr) (Richardson *et al.*, 2009). Chronic diarrhoea without pyrexia, sometimes intermittent, body weight loss, BCS loss. Appetite remains normal. Decrease of milk production and increase of somatic cell count in lactating animals. Those animals will become thin with a poor body condition. A cachectic syndrome is common.
- Risks of infection decrease with time during the first year of age; animals over 1 year old have an age-resistance against the pathogen.
- Recent evidence suggests horizontal transmission in older heifers, and even in cows.

Diagnosis:
- Faeces culture, PCR, ELISA, agar gel immunodifusion (AGID). The golden standard for Map diagnosis is the faeces culture, but this is highly time-consuming. If one detects a single animal with the ELISA test and the culture of faeces is positive, this means that much more infected animals are present in the herd, or that the test-positive animal was a purchased one.

---

[5] Adapted after Baumgartner, 2006; Barrett *et al.*, 2008; Radostits *et al.*, 2007; Richardson *et al.*, 2009; Smith, 2009

Risk factors:
- Open herds, large herds, old herd age profile, communal calving facilities, feeding pooled colostrum, feeding unpasteurised waste milk, management of rearing female calves, herd depopulation, contacts among animals and with humans, hygiene and disinfection, purchase, geographic region, hygiene, handling and storage of manure, access of wild animals to the feed storage, and quality of water, among others (after Barrett *et al.*, 2008).

Strategy:
- No treatment is available. Best practice is: (a) controlling the infection of newborn calves and young stock up to one year of age, (b) detection and culling of infected animals, including the last calf born from a clinically diseased dam. Other measures are: cleaning the teats of the dam before colostrum milking, clean and disinfect calving pen, milking colostrum aseptically and giving it with a teat-bottle. Remove the calf as quickly as possible from its mother. Calves are infected via e.g. colostrum: the pasteurisation of colostrum can reduce the risk of this transmission. Feeding of silage or hay from pasture plots where no paratuberculosis-positive manure has been spread.

Control:
- Identify and cull clinical cases. Install strict biosecurity plans. Clean facilities regularly and thoroughly (no manure rests must be present).
- Eradication is highly costly and takes a long time (see Chapter 4 on biosecurity in Section I). Controlling the disease, with biosecurity, and culling of clinically affected animals, can reduce the risk of transmission.
- Biosecurity plans must be developed. The farm faeces must be taken out of the barns as quickly as possible, contact between younger calves and older animals must be avoided (this includes the use of separate boots and clothes too), and different pastures according to the age must be assigned. Control of environmental conditions, properly handling the manure and the avoidance of contact of paratuberculosis with calves under one year of age are a few key elements in this biosecurity.
- The Crohn's disease in humans seems somehow related with Map, but evidence of a sole agent causal link is currently inconclusive.

Vaccination:
- A killed vaccine is available but will interfere with the Tuberculosis test. It reduces clinical cases, but it is not possible to eradicate the disease only with vaccination!
- Herd management (preventive measures and biosecurity) is the key point for controlling this disease.

13.9. Q-fever

- *Occurrence*: high prevalence in ruminants; latent infection, and excretion at parturition. Direct contact and via inhalation. Very important zoonotic disease, worldwide spread.
- *Causal agent*: *Coxiella burnetii*.
- *Signs*: at any age: anorexia, usually not apparent.
- *Diagnosis*: difficult to diagnose from clinical signs alone.
- ELISA; indirect immuno-fluorescence; complement fixation; micro agglutination, PCR and RT-PCR in the milk.
- *Treatment and prevention*: vaccination with killed vaccines.

13.10. Ringworm (*Trichophyton verrucosum*)

- *Occurrence*: young stock over winter in high stocking density housing, fungal spores survive for years in housing, particularly in wooden fittings, immunity develops after infection.
- *Signs*: circular partial alopecia followed by grey dry crust formation on young stock and erythematous circular lesions on the hands, arms and head of humans attending affected stock.
- *Diagnosis*: classical lesions and distribution on body.
- *Treatment*: in-feed or topical anti-mycotic medication, self-cure upon turnout to pasture in well-fed stock, need to treat in-contacts as initial phase of lesion development is inapparent.
- *Prophylaxis*: disinfection of the young stock housing, prevention of comingling of young stock with older stock, vaccination, wear protective clothing when handling affected stock. Infection has a zoonotic character: children and elderly people are most susceptible.

13.11. Urolithiasis

- *Occurrence*: housed calves with milk substitute, pastures in problem areas, high plant oestrogen, oxalate or silica levels, calculi in the males in sigmoid flexure.
- *Causes*: urinary calculi are either organic (less common) or inorganic, calculi in housed animals – Ca or magnesium ammonium phosphate, pasture: carbonates of Ca, magnesium and phosphorus, more frequent in castrated animals.
- *Signs*: partial or total blockage of the urethra, calculi at the prepucial orifice hairs, mild or severe colic, kicking and bellying, if there is a rupture then septicaemia can be expected.
- *Diagnosis*: history, area of feeding, sex, urinary crystals may be present on prepucial hairs; differential diagnosis includes ascites, intussusception and constipation.

- *Treatment and prevention*: surgery; no treatment → slaughter; treatment with success only in early stages: urethrotomy and remove the calculi. Medical- Hyoscine butylbrumide IV or IM at a dose of 20-40 mg/animal. Prevention- feed changes. Correct ratio Calcium-to-Phosphor, low Mg levels in the diet, urinary acidifiers like ammonium sulphate or phosphoric acid, in pasture supply salt to reduce the silica acid concentration in the urine.

13.12. Vitamins and trace elements

Calcium, Phosphorus and vitamin D deficiencies

- *Calcium deficiency* when high levels of cereals are fed, so often in the best growing calves. The diagnosis is based on clinical signs like stiffness, fracture of long bones, poor growth and dentation. Daily requirements are 10 to 30 g Ca.
- *Phosphorus deficiency* can occur with low vitamin D levels in the ration, high Ca or vitamin A levels, excessive Ca feeding, on particular pasture plots or soils, when Fe and Al levels are high.
- *Vitamin D deficiency* can be seen when poor quality hay is fed. The back of the animal is arched. The daily requirements are 7-12 IU/kg bodyweight.

Copper deficiency (Cu)

- *Occurrence*: few old calves, normally 2-3 months of age.
- *Causes*: lack of copper in the diet or secondary deficiency due to a dietary antagonist most commonly molybdenum, failure of digestion, low copper in (cow)milk ~~ see Annex.
- *Signs*: skeletal defects, illthrift and possibly poor coat colour.
- *Treatment*: oral administration of 1.5 g copper sulphate.
- *Prevention*: assure that levels of copper in the diet of dams and calves are at least 10 mg/kg bodyweight.

Iodine deficiency

- *Occurrence*: new born calf often death, abortions.
- *Causes*: primary lack of iodine, high intake of brassica, high calcium ingestion, low level intake of linseed meal, areas with high rainfall, Iodine forms part of the hormone thyroxin – necessary for metabolism and respiratory function. Iodine deficiency interacts with selenium deficiency.
- *Signs*: new born calves with thyroid enlargement, alopecia areas very rare, weak live calves, which do not suck.

- *Diagnosis*: thyroid enlargement (goitre), abortion of several heifers or cows or stillborn calves, very week calves.
- *Treatment and prevention*: ensure that the dam has an adequate intake of iodine and of selenium during pregnancy. Assure that calf sucks, dry and draught-free environment, intravenous sodium iodine injections, dose: 1-2 mg per kg body weight.

Iron (Fe) deficiency

- *Occurrence*: in veal calves or others fed predominantly raw cow milk or unsupplemented milk substitute The calf has only sufficient hepatic iron reserves for about 3 weeks after birth, only source is milk substitute.
- *Causes*: veal calves without access to roughage, after haemorrhages, heavy infestation with lice (*Haemathopinus eurysternus* and *Linognathus vituli*).
- *Signs*: reduction of appetite and poor weight gain, mucous membranes, pale, death is extremely rare, moderate anaemia.
- *Diagnosis*: necropsy – pale muscles – blood is thin and watery and clotting slowly, enlarged liver; diagnosis depends on the history of the diet and signs.
- *Treatment and prevention*: 1 injection of 1 g of Iron on a weekly basis.

Selenium (Se) and vitamin E deficiency

- *Occurrence*: at any age after birth, also in adult animals, affects muscles (cardiac, skeletal and diaphragmatic), suckler calves sucking dams with low Se or vitamin E levels.
- *Causes*: deficiency in vitamin E and Se, soils poor in Se, vitamin E – type of crop grown and its storage, oxidation of calves diets very rich in unsaturated fatty acids with rancidity and destruction of vitamin E.
- *Signs*: muscular dystrophy – white muscle disease, sudden death syndrome within a minute of collapse, lateral recumbence, respiratory distress, heart rate irregular and 150-200 beats per minute, rectal temperature normal; in adult animal related with infertility.
- *Diagnosis*: weak calf and will not stand for long, remains conscious with normal appetite, affected muscles could be swollen, increased risk of pneumonia, response to treatment, blood analysis (see Table 13.3).
- *Treatment and prevention*: application of vitamin E and or Se, dose: D2 – Alfa tocopherol is about 6 IU/kg BW, as prevention supplementation, supplement of vitamin E in the last 60-45 days before calving, add Se to drinking water or apply as a slow release rumen bolus. Reference values for vitamin E for both the individual animal and groups are given in Table 13.3.

Table 13.3. Reference values for vitamin E in blood (in mmol alpha-tocopherol/l):

Individual animal	Pooled sample (n=5)	Interpretation
<4	<4.9	deficient
4.0-7.4	4.9-7.9	suboptimal
>7.4	>7.9	sufficient

Zinc (Zn) deficiency

- *Occurrence*: in calves from 6 to 10 weeks of age, period after weaning and normally housed calves.
- *Causes*: primary due to a lack of Zinc, secondary due to impaired intake, congenital form in Friesian calves.
- *Signs*: 2 weeks after the introduction of the Zn-deficient diet, poor growth, alopecia, parakeratosis, wounds will take longer to heal.
- *Diagnosis*: biopsy shows parakeratosis, high serum globulins and serum alkaline phosphatase, low albumin and amylase levels.
- *Treatment and prevention*: oral administration of zinc sulphate – 2 g weekly, diets containing 50 ppm.

Chapter 14. Tropical and subtropical diseases in young stock[6]

14.1. Anthrax

- *Occurrence*: highly contagious disease; grazing contaminated soil and or animal by-products.
- Worldwide spread, mainly in Africa and Asia, United States, and also in some Mediterranean countries.
- *Causal agent*: *Bacillus anthracis*
- *Signs*: peracute disease with sudden death. Bloody exudates from body cavities.
- *Diagnosis*: anamnesis, clinical signs, contagiosity. Differential diagnostics with: plant poisoning, mineral deficiencies. Examination of direct blood or tissue smears and bacterial culture (intact eye) – samples from lymph nodes, fluids.
- *Treatment and prevention*: according to the OIE this is an immediately notifyable disease. No treatment!!
- Destruction of the carcass must be done. Vaccination with a live acapsular vaccine strain.
- Important note: do not open the carcass!! This is a zoonotic disease!!

14.2. Besnoitiosis (Elephant skin disease)

- *Occurrence*: endemic disease in some tropical and subtropical areas with high rate of morbidity and low mortality; in other areas a rare disease.
- *Causal agents*: *Besnoitia besnoti*. They are coccidian parasites in the family *Sarcocystidae*.
- *Signs*: anasarca; alopecia; hyperpigmentation and scleroderma; loss of condition, inspiratory dyspnea; pin point nodules on the scleral, conjunctival, nasal, pharyngeal and laryngeal mucosa. Parasitic cysts in dermis, subcutaneously.
- *Diagnosis*: clinical signs. Parasitic cysts in dermis, subcutaneously. Fever, painful swellings, generalized edema of the skin. Superficial lymph nodes are swollen. Diarrhoea may occur. Lacrimation and nasal discharge.
- In the chronic stage the skin becomes grossly thickened, corrugated and alopecia. Cysts on the scleral conjunctiva are considered to be a important part of the diagnostic process. Demonstration of *Besnoitia bradyzoites* in skin biopsy smears or scleral conjunctival scrapings.
- *Treatment and prevention*: not available. Vaccination might be an option.

---

[6] Text summarised from and adapted after Dirksen *et al.* (2000), Radostits (2000), Radostits *et al.* (2007), OIE (2004), Divers and Peek (2008), Smith (2009). Diseases with any relevance for HHPM are retained.

14.3. Contagious Bovine Pleuropneumonia, CBPP

- *Occurrence*: rarely present in Europe; mainly in Africa, and South America in endemic areas.
- *Causal agent*: *Mycoplasma mycoides*.
- *Signs*: severe pneumonia but can also evolve without evident clinical signs.
- *Diagnosis*: severe pneumonia, not reactive to any kind of treatment. Blood samples and complement fixation test. Differential diagnosis with MCF, BRD, and other pneumonia.
- *Treatment and prevention*: No treatment!! Slaughter all positive animals; transport control.
- In infected areas: restrict animal movements, apply biosecurity measures; close the area.

14.4. Foot-and-mouth disease, FMD

- *Occurrence*: all over the world; highly contagious disease with low mortality. Large economic impact.
- *Causal agent*: *Apthovirus*.
- *Signs*: fever, excess salivation, vesicles in mouth, feet. Sudden death in young animals.
- Erosive – ulcerative stomatitis and oesophagitis.
- In neonates – interstitial mononuclear and necrotic myocarditis.
- *Diagnosis*: blood samples, virus isolation, serology, RT-PCR detection.
- *Treatment and prevention*: no treatment, only symptomatic with some good results reported.
- Mass vaccination with killed vaccines in endemic areas.
- *Eradication* by slaughter. Different types of eradication campaigns exist (with and without (ring) vaccination and buffer zones.

14.5. Haemoparasitic diseases

14.5.1. Anaplasmosis (Rickettsiosis)

- *Occurrence*: in calves in the first 6 to 9 months of life, but much more severe in older animals. Incubation time 15 to 30 days after tick infestation; normally in enzootic areas; widely spread in Southern Europe, Mediterranean countries, South and Central America, UK, Africa.
- *Causal agents*: protozoan parasites: *Anaplasma marginale, Anaplasma centrale*.
- *Signs*: progressive anemia; subclinical infection; anorexia; in adult cattle high fever and if no treatment, decreasing temperature and death.

- *Diagnosis*: icterus, fever, depression, anemia, severe dyspnea. Sometimes it is possible to find ticks (*Boophilus* spp., *Ixodes ricinus, Dermacentor andersoni, D. variabilis, Rhipicephalus* species) and flies.
- Peripheral blood samples, and blood smears on Giemsa or Elisa tests, PCR. Hystology – fixed spleen, liver, bone marrow.
- *Treatment and prevention*: tetracyclines 11 mg/kg BW IV or long acting oxytetracycline at 20 mg/kg BW IM at 72 hours interval; blood transfusion as supportive therapy; Diminazine aceturate at 3 to 5 mg/kg; Phenamidine diisethionate at 8 to 13 mg/kg; Imidocarb diprionate at 1 to 3 mg/kg, or Amicarbalide diisethionate at 5 to 10 mg/kg. Baths or spraying with Amitraz; Permethrin products.

14.5.2. Babesiosis

- *Occurrence*: 2 to 3 weeks after tick infestation; normally in enzootic areas; widely spread in Southern Europe, Mediterranean countries, South and Central America, British Isles, Africa.
- *Causal agents*: Protozoan parasites; *Babesia bigemina; B. bovis (B. argentina, B. berbera); B. divergens; B. major; B. jakimov; B. ovata.*
- *Signs*: fever up to 40 °C-42 °C; depression, icterus, anorexia, tachycardia, tachypnea, anemia, haemoglobinemia, haemoglobinuria, abortion, and death. Cerebral babesiosis with hyperexcitability, convulsions, opisthotonus, coma, and death.
- *Diagnosis*: clinical signs, icterus, fever, depression, anemia. Sometimes it is possible to find ticks (*Boophilus annulatus, B. decoloratus, B. microphus, B. microplus, Ixodes ricinus,*
- *Haemaphysalis punctata, Haemaphysalis longicornis, Dermacentor, Hyalomma, Rhipicephalus* species, *Haemaphysalis* species). Peripheral blood samples, and blood smears on Giemsa.
- *Treatment and prevention*: Boviprovicona; Tetracyclines; blood transfusion; Diminazine aceturate at a dose of 3 to 5 mg/kg BW; Phenamidine diisethionate at 8 to 13 mg/kg BW; Imidocarb diprionate at 1 to 3 mg/kg, or Amicarbalide diisethionate at 5 to 10 mg/kg BW.
- Vaccination with live vaccine. Baths or spraying with Amitraz; Permethrin products.

14.5.3. Theileriosis (East Coast Fever, ECF)

- *Occurrence*: endemic disease mainly in east and central Africa; also widely spread in Southern Europe, Mediterranean countries, South and Central America.
- *Causal agents*: protozoan parasites: *Theileria parva, T. taurotragi, T. annulata, T. buffeli, T. Mutans, T. sergenti.*

- *Signs*: fever; enlarged superficial lymph nodes; jaundice; dyspnea; wasting and terminal diarrhoea.
- *Diagnosis*: icterus, fever, depression, anemia, severe dyspnea. Sometimes it is possible to find ticks (*Rhipicephalus apendiculatus, Rhipicephalus* spp., *Hyalomma anatolicum, Haemophysalis* spp., *Amblyomma* spp., *H. longicornis, H. punctata, Haemophysalis* spp.). Peripheral blood samples, and blood smears on Giemsa.
- *Treatment and prevention*: limited success with halofuginone at 1.2 mg /kg BW; Buvaparvoquone 2.5 mg/kg BW, or Tetracyclines at 20 mg/kg BW. Aplication of acaricides and vaccination. Baths or spraying with Amitraz; Permethrin products.

14.6. Rinderpest

- *Occurrence*: affects ruminants in Asia, Middle East, Africa. Highly contagious and high mortality. Cattle of all ages are animals at risk.
- *Causal agents*: *Morbillivirus*. Family *Paramyxoviridae*. Many strains.
- *Signs*: hyperthermia, ocular and nasal discharge, salivation, ulcerative stomatitis, diarrhoea – severe and dysentery with tenesmus. Skin lesions affecting perineum, later on with scabs, oesophagitis, dehydration and dead. Could be peracute, acute, subacute or innapparent in other species. Discrete lesions, greyish, raised necrotic lesions, first in the lower lip (1-5 mm) and adjacent gum or in the tongue.
- *Diagnosis*: necrotic stomatitis, oesophagitis, enterocolitis (ulcerative and haemorrhagic), massive necrosis of lymphnoytes in Peyer's patches, lymph nodes and spleen. Marked leucopenia. Virus isolation; serology; immuno – histochemistry; Agar gel diffusion (AGID). Differential diagnosis: BVDV, FMD; MCF.
- *Treatment and prevention*: no treatment!! Apply strict biosecurity measures.

References

Andrews, A.H, Blowey, R.W, Boyd, H., Eddy, R. 1992. Bovine Medicine, Diseases and Husbandry of Cattle. Blackwell Science, London, UK, pp. 231-252.

Andrews, A.H, Blowey, R.W., Boyd, H., Eddy, R.G. 2008. Medicina Bovina, segunda edição, editado por Editora Ltda Roca, Sao Paulo, Brasil.

Anonymous, 2009. Outbreaks of idiopathic haemorrhagic diathesis syndrome in young calves. Veterinary Record 165: 139-142.

Argyle, M. 1994. Face, gaze and other non-verbal communication. In: The psychology of interpersonal behaviour, 5th ed., Penguin Psychology, London, UK, pp. 23-55.

Arrow, K.J. 1996. The theory of risk-bearing: small and great risks. Journal of Risk and Uncertainty 12: 103-111.

Barrett, D., Mee, J.F., Good, M., Mullowney, P., Cleg, T., More, S. 2008. Risk factors for Johne's disease in Irish dairy herds: A case-control study. Cattle Practice 6: 209-215.

Barry, P.J., Ellinger, P.N., Hopkin J.A., Baker C.B. 2000. Financial management in agriculture. Interstate Publishers, Danville, IL, USA.

Baumgartner, W. 1999. Klinische Propaedeutik der inneren Krankheiten und Hautkrankheiten der Haus – und Heimtiere. Parey Buchverlag, Berlin, Germany.

Baumgartner, W. 2006. What are we doing against paratuberculosis in Austrian cattle? Document of the UVG Conference, Coimbra, Portugal, 4th July 2006.

Bergevoet, R.H.M. 2005. Entrepreneurship of Dutch dairy farmers. PhD thesis University of Wageningen, Dept. of Farm Management, Wageningen, the Netherlands.

Bertrand, M., Mullainathan, S., Shafir, E. 2007. Behavioral economics and marketing in aid of decision-making among the poor. Journal of Public Policy and Marketing 25(1): 8-23.

Bexiga, R., Guyot, H., Saegerman, C. 2009. Differential diagnosis of bluetongue. In: Blue Tongue in Northern Europe (Saegerman, Reviego-Gordejo and Pastoret, editors). Edition of the OIE, Paris, France, pp. 61-71.

Boersema, J.S.C. 2006. Development of a HACCP-compatible programme in aid to quality risk management on dairy farms (with emphasis on the rearing of the dairy replacement period). MSc thesis Utrecht University, August 2006, Utrecht, the Netherlands.

Boersema, J.S.C., Noordhuizen, J.P.T.M., Vieira, A., Lievaart, J.J., Baumgartner, W. 2008. Imbedding HACCP principles in Dairy Herd Health and Production Management: case report on calf rearing. Irish Veterinary Journal 61(9): 594-602.

Boersema, J.S.C., Noordhuizen, J.P.T.M. 2010. Applied biosecurity on dairy farms. Bulletin of the SNGTV. SNGTV, Paris, France (in press, in French).

Bouwknegt, M., Dam-Deisz, W.D.C., Wannet, W.J.B, Van Pelt, W., Visser, G., Van der Giessen, A.W. 2004. Surveilance of zoonotic bacteria in farm animals in the Netherlands. Report 330050001/2004. State Institute of Public Health and Environment (RIVM), Bilthoven, the Netherlands.

Bracke, M.B.M., Spruijt, B.M., Metz, J.H.M., Schouten, W.G.P. 2002. Decision support system for overall welfare assessment in pregnant sows: model structure and weighing procedure. Journal of Animal Science 8: 1819-1834.

Bramley, A.J., Dodd, F.H. 1984. Reviews of the progress of dairy science: mastitis control – progress and prospects. Journal of Dairy Research 51: 481-512.

Brand, A., Noordhuizen, J.P.T.M., Schukken, Y.H. 1996. Herd health and production management in dairy practice. Wageningen Academic Publishers, Wageningen, the Netherlands.

Brandt, J.C., Batemann, S.W. 2006. Senior Veterinary Students' Perceptions of Using Role Play to Learn Communication Skills. Journal of Veterinary Medical Education 33 (1): 76-80

Brickell, J.S., McGowan, M.M., Pfeiffer, D.U., Wathes, D.C. 2009. Mortality in Holstein-Friesian calves and replacement heifers, in relation to body weight and IGF-I concentration, on 19 farms in England. Animal 3:1175-1182.

CAC, Codex Alimentarius Commission. 2003. General principles of food hygiene (CAC/RCP 1-1969, Rev. 4-2003), incorporating the hazard analysis critical control point (HACCP) system and guidelines for its application. WHO and FAO, Rome. Available at http://www.codexalimentarius.net/web/standard_list.do?lang=eng.

Cady, R.A., Smith, T.R. 1996. Economics of heifer raising programs. In: Proc. Calves, Heifers and Dairy profitability. National Congress Harrisburg PA. NRAES publ.74, Ithaca, NY, USA.

Cannas da Silva, J. 2002a. Controlo do IBR i BVD (*the control of IBR and BVD*). Congresso de Reprodução Animal, Porto, Portugal, pp. 3-9.

Cannas da Silva, J. 2002b. Perspectivas e métodos para controlo do BVD (*perspectives and control methods of BVDV*). In: Proceedings of the IIId Congresso Ibérico de Reprodução Animal, pp. 65-69.

Cannas da Silva, J., Noordhuizen, J.P.T.M, Vagneur, M., Bexiga, R., Gelfert, C.C., Baumgartner, W. 2006. Veterinary dairy herd health management in Europe: constraints and perspectives. Veterinary Quarterly 28(1): 23-32.

Chassange, M., Barnouin, J., Charconac, J.P. 1999. Risk factors for stillbirth in Holstein heifers under field conditions in France: a prospective study. Theriogenology 51: 1477-1488

Cockcroft, P., Holmes, M. 2003. Evidence-based medicine: What is it? Why is it important? What skills do I need? Cattle Practice (BCVA) vol. II part 4, pp. 373-384.

Collell Verdaguer, J. 2007. Reproduccion en vacuno lechero. Cuadernos de Campo, Merial Europe, pp. 114, 125.

Collins, M.T., Morgan I.R. 1992. Economic decision analysis model of a paratuberculosis test-and-cull program. Journal of the American Veterinary Medical Association 199(12): 1724-1729.

Correa, M.T., Erb, H., Scarlett, J. 1993. Path analysis for seven post partum disorders of Holstein cows. Journal of Dairy Science 76: 1305-1312.

Cullor, J.S. 1998. HACCP: is it coming to the dairy? Journal of Dairy Science 80: 3449-3452

De Clerq, K., Vandenbussche, F., Van Binst, T., Vandenmeulebroucke, E., Goris, N., Zienstra, S. 2009. Blue tongue: laboratory diagnosis (chapter 9). In: Blue Tongue in Northern Europe (Saegerman, Reviego-Gordejo and Pastoret, editors). Edition of the OIE, Paris, France, pp. 72-83.

De Kruif, A., Mansfeld, R., Hoedemaker, M. 2007. Tierartzliche Herdenbetreuung beim Milchrind. 2nd Edition. Enke Verlag, Stuttgart, Germany.

Dirksen, G., Gruender, H.D., Stoeber, M. 1984. Innere Medizin und Chirurgie des Rindes. Parey Buchverlag MVS, Stuttgart, Germany.

Dirksen, G., Gruender, H.D., Stoeber, M. 2005. Medicina Interna y Cirurgia del Bovino, 4ª Ediçao. Inter-Médica XXI-2005, Buenos Aires, Argentina.

Divers, T.H.J., Peek, S.F. 2008. Rebhun's Diseases of Dairy Cattle. 2nd Edition. W.B. Saunders/ Elsevier, St. Louis, MO, USA.

Dutch Dairy Manual. 2006. Handboek Rundveehouderij, WUR, Lelystad, the Netherlands.

DQA. 2000. Biosecurity guidelines (Ruegg, P. chief-editor). Dairy Quality Assurance, the Milk and Dairy Beef Quality Assurance Center Inc., Stratford IA, USA.

FAO, Food and Agriculture Organisation of the United Nations. 1998. A training manual on food hygiene and the hazard analysis and critical control points (HACCP) system. FAO, Rome (I). Available at http://www.fao.org/documents/index.asp.

FAO, Food and Agriculture Organisation of the United Nations. 2004. Guide to good dairy farming practice. FAO, Rome, Italy.

FAWC. 1992. Farm Animal Welfare Council: FAWC updates the five freedoms. Veterinary Record 131: 357.

Fetrow, J. 1985. Economic decisions in veterinary practice: a method for field use. Journal of the American Veterinary Medical Association 186 (8): 792-797.

Foley, J,A., Otterby, D.E. 1978. Availability, storage, treatment, composition and feeding value of surplus colostrum: a review. Journal of Dairy Science 61: 1033-1060.

Franken, P. 2009a. IBR control: the Dutch approach. Symposium LMV Lda, 27th March 2009, Evora, Portugal. Proceedings available at www.lmv.com.pt.

Franken, P. 2009b. Bovine Virus Diarrhoea control: the Dutch approach. Symposium LMV Lda, 27th March 2009, Evora, Portugal. Proceedings available at www.lmv.com.pt.

Fridrich, A., Rademacher, G., Weber, B.K., Kappe, E., Carlin, A., Assad, A., Sauter-Louis, C., Hafner-Marx. A., Buettner, M., Boucher, J., Klee, W. 2009. Gehaueftes Auftreten von Haemorrhagischer Diathese infolge Knochenmarkschaedigung bei jungen Kaelber. Tierartztl. Umschau, 64: 423-431.

Gardner, I.A., 1997. Testing to fulfill HACCP (hazard analysis critical control points) requirements: principles and examples. Journal of Dairy Science 80(12): 3453-3457.

Gelfert, C.C., Staufenbiel, R., 2004. Early detection of metabolic diseases of dairy cattle by using milk data, body condition and metabolic profiles. In: Proceedings of the 23rd World Buiatrics Congress, Québec, Canada, 11-16 July 2004, p. 82.

Godden, S.M., Fetrow, J.P., Feirtag, J.M., Green, L.R., Scott, M.S., Wells, J. 2005. Economic analysis of feeding pasteurized nonsaleable milk versus conventional milk replacer to dairy calves. Journal of the American Veterinary Medical Association 226(9): 1547-1554.

Godden, S.M., Haines, D.M., Konkol, K., Peterson, J. 2009. Improving passive transfer of immunoglobulins in calves. II: Interaction between feeding method and volume of colostrum fed. Journal of Dairy Science 92: 1758-1764.

Groenendaal, H., Nielen, M., Jalvingh, A.W., Horst, H.S., Galligan, D.T., Hesselink, J.W. 2002. A simulation of Johne's disease control. Preventive Veterinary Medicine 54(3): 225-245.

Groenendaal, H., Galligan, D.T. 2003. Economic consequences of control programs for paratuberculosis in midsize dairy farms in the US. Journal of the American Veterinary Medical Association 223(12): 1757-1763.

Grunert, E. 1984. Buiatrik, Band I, Verlag Schaper, Hannover, Germany.

Hardaker, J.B., Huirne, R.B.M., Anderson, J.R. 1997. Coping with risk in agriculture. CAB International, Wallingford, UK.

Harrington, S.E., Niehaus, G.R. 1999. Risk management and insurance. The McGraw-Hill Companies, Boston, USA.

Heuchel, V., Parguel, P., David, V., Lenormand, M., Le Mens, P. 1999. Maitrise de la qualité hygiénique en production laitière: l'application d'HACCP en élevage. In: Proceedings du Rencontres de Recherches en Ruminants, pp. 291-297.

Howard, J.L. 1993. Current Veterinary Therapy 3, Food Animal Practice. W.B. Saunders Company, Philadelphia, PA, USA.

Huirne, R.B.M., Saatkamp, H.W., Bergevoet, R.H.M. 2002. Economic analysis of common health problems in dairy cattle. In: Proceedings of the XXIIth World Buiatrics Congress (Kaske, Scholz and Holterschinken, editors), 18-23 August 2002, Hannover, Germany, pp. 420-431.

Heinrichs, J., Kehoe, S.I. 2009. Electrolytes – oral rehydration solutions for scouring calves. International Dairy Topics 8(5): 13-15.

Heuvelink, A.E., Van den Biggelaar, F.L., Zwartkruis-Nahuis, J.T.M., Herbes, R.G, Huyben, R., Nagelkerke, N., Melchers, W.J.G., Monnens, L.A.H., De Boer, E. 1998. Occurence of verocytotoxin producing *Eschirichia coli* O_{157} on Dutch dairy farms. Journal of Clinical Microbiology 36: 3480-3487.

Heuvelink, A.E., Van Heerwaarden, C., Zwartkruis-Nahuis, J.T.M., Van Oosterom, R., Edink, K., Van Duynhoven, Y.T.H.P., De Boer, E. 2002. *Escheria Ecoli* O_{157} infection associated with a petting zoo. Epidemiology and Infection 129: 295-302.

IKC. 1994. Quality control in dairy farming. IKC Lelystad and Landbouwschap, The Hague, the Netherlands (in Dutch).

Intervet. 2006. Compendium of Animal Reproduction. Intervet International, Boxmeer, the Netherlands, p. 18.

Jimeno Vinatea, J.V. 2009. Alimentacion de novillas de reposicion. EUIT Agricola, Madrid, Spain. Postgraduate course, Universidade Lusofona, Lisbon, Portugal, July 4th and 5th 2009.

Juaristi, J.L., Bach, A., Rodriguez, P. 2007. Cow comfort. Cuadernos de Campo, Merial Europe.

Kappe, E.C., Halami, M., Schade, B., Bauer, J. 2009. Fatal aplastic anaemia with haemorrhagic disease in calves in Germany. Proceedings of the 27th Meeting of the European Society of Veterinary Pathologists and European College of Veterinary Pathologists. 9-12 September, 2009, Olsztyn-Krakow, Poland, p.118.

Karmali, R.A., Marsh, J. and Fuchs, C. 1985. Effects of dietary enrichment with gamma-linolenic acid upon growth of the R3230AC mammary adenocarcinoma. Journal of Nutrition and Growth Cancer 2: 41-51.

Kelton, D.F., Lissemore, K.D., Martin, R.E. 1998. Recommendations for recording and calculating the incidence of selected clinical diseases of dairy cattle. Journal of Dairy Science 81: 2502-2509.

Kleen, J.L. 2008. Communication for veterinary advisory tasks on large dairy farms. In: Proceedings of the ALLTECH and VACQA-International Workshop on Veterinary Management on Large Dairy Farms (Noordhuizen and Andrieu, editors), at the XXVth World Buiatrics Congress, Budapest, Hungary, 10th July 2008, pp. 31-34.

Kleen, J.L., Rehage, J. 2008. Communication skills in veterinary medicine. Tierarztliche Praxis 36: 293-297.

Krebs, S., Danuser, J., Audigé, L., Kihm, U. 1999. Evaluation eines Monitoringsystems zur Erfassung der Tiergesundheit beim Milchvieh. Schweiz. Archives Tierheilkunde 141: 559-565.

Kunz, H.J. 2008. Milk replacers for dairy young stock. DSM Annual Meeting on Ruminants, Poitiers (F), March 18th 2008, DSM, Paris, France.

Lievaart, J.J., Noordhuizen, J.P.T.M., Van Beek, E., Van der Beek, C., Van Risp, A., Schenkel, J., Van Veersen J. 2005. The hazard analysis critical control points concept as applied to some chemical, physical and microbiological contaminants of milk on dairy farms. The Veterinary Quarterly 27(1): 21-29.

Martins, S., Sousa, S., Madeira de Carvalho, L.M., Bacelar, J., Cannas da Silva, J. 2007. Prevalence of *Cryptosporidium parvum* infection in Northwest Portugal dairy calves and efficacy of halofuginon lactate on the prevention of cryptosporidiosis. BCVA Cattle Practice vol.15, part 2, pp. 152-156.

Martins, S., Madeira de Carvalho, L.M., Cannas da Silva, J. 2007. *Cryptosporidium parvum*: major aetiological agent of neonatal diarrhoea in dairy calves. In: Proceedings of the 21st International Conference of the WAAVP 'From EPG to Genes', Ghent, Belgium, 19-23 August 2007, p.31.

McFadden, D. 1999. Rationality for economists? Journal of Risks and Uncertainty 19 (1-3) 73-105.

McGuirk, B.J., Forsyte, R., Dobson, H. 2007. Economic cost of difficult calvings in the United Kingdom dairy herd. Veterinary Record 161(20): 685-687.

Mee J.F., Mehra, R. 1995. Efficacy of colostrum substitutes and supplements in farm animals. Agro-Food-Industry Hi-Tech 6: 31-35.

Mee, J.F., Wafa, S., O'Farrell, K.J. 1995. Clinical response and economic value of vaccinating dairy calves with two types of vaccines. Agri-Practice 16: 15-19.

Mee, J.F., O'Farrell, K.J., Reitsma, P., Mehra, R. 1996. Effect of a whey protein concentrate as a colostrum substitute or supplement on calf immunity, weight gain and health. Journal of Dairy Science 79: 886-894.

Mee, J.F. 2004. The role of micronutrients in bovine periparturient problems. In: Proceedings of the Cattle Association of Veterinary Ireland Annual Conference, Bunratty, Co. Clare, pp. 39-56.

Mee, J.F. 2007. The role of the veterinarian in bovine fertility management on modern dairy farms. Theriogenology, 68S: 257-265.

Mee, J.F. 2008a. Newborn dairy calf management. Veterinary Clinics of North America Food Animal Practice, 24: 1-17.

Mee, J.F. 2008b. Managing the cow at calving time. Proceedings of the American Association of Bovine Practitioners 41: 35-43.

Mee, J.F. 2008c. Prevalence and risk factors for dystocia in dairy cattle: A review. The Veterinary Journal 176: 93-101.

Mee, J.F. 2008d. Perinatal mortality in heifers – An emerging problem. Cattle Practice 16: 166-173.

Mee, J.F. 2008e. Stillbirth in dairy cattle – From science to solutions. In: Proceedings of the XXV Jubilee World Buiatrics Congress, Budapest, Hungary, pp. 82-86.

Mee, J.F., Berry DP, Cromie AR. 2008. Prevalence of, and risk factors associated with, perinatal calf mortality in pasture-based Holstein-Friesian cows. Animal 2: 613-620.

Mee, J.F. 2009. Bovine perinatology: Current understanding and future developments. In: Animal Reproduction: New Research Developments (L.T. Dahnof, editor). Nova Science Publishers, Inc. New York, pp. 67-106.

Mestdagh, C., Raboisson, D. Schelcher, F. 2008. Colostrum substitutes: how to use them. Le Nouveau Praticien Vétérinaire 167: 19-26 (in French).

Metz, J.H.M., Bracke, M.B.M. 2003. Assessment of the impact of locomotion on animal welfare. In: Abstracts of the EAAP annual meeting, Rome, Italy. Wageningen Academic Publishers, Wageningen, The Netherlands, p. 212.

Morrow, D.A. 1986. Current therapy in Theriogenology 2. W.B. Saunders Company, Philadelphia, PA, USA, pp. 414-422.

Noakes, D.E., Parkinson, T.J., England G.C.W. 2001. Arthur's Veterinary Reproduction and Obstetrics. Saunders/Elsevier Science Ltd, St. Louis, MO, USA.

Noordhuizen, J.P.T.M., Brand, A., Dobbelaar, P. 1983. Veterinary herd health and production control on dairy farms. I: introduction to a coupled basic system and flexible system. Preventive Veterinary Medicine 1: 189-199.

Noordhuizen, J.P.T.M., Welpelo, H.J. 1996.Sustainable improvement of animal health care by systematic Quality Risk Management according to the HACCP concept. The Veterinary Quarterly 18: 121-126.

Noordhuizen, J.P.T.M. 2004. Dairy Herd Health and Production Management practice in Europe: state of the art. In Proceedings of the 23rd World Buiatrics Congress, Québec, Canada, 11-16 July 2004.

Noordhuizen, J.P.T.M., Lievaart, J.J. 2005. Cattle welfare and Cow comfort. In: Proceedings of the 1st Buiatrissima Congress, Bern, Switzerland, pp. 1-12.

Noordhuizen, J.P.T.M., Cannas da Silva, J., Boersema, J.S.C., Vieira, A. 2008. Applying HACCP-based quality risk management on dairy farms. Wageningen Academic Publishers, Wageningen, the Netherlands.

NRC, National Research Council. 2001. Nutrient requirements of dairy cattle. 7th Revised edition. The National Academies Press, Washington DC, USA.

Nuotio, L., Neuvonen, E., Hyytiainen, M. 2007. Epidemiology and eradication of IBR/IPV virus in Finland. Acta Veterinaria 49: 3.

OIE, Office Internationale d'Epizooties. 2004. Blue tongue, updated 22/4/2002. - Classified as an OIE *List A* disease (A090). Available at www.oie.int/Eng/maladies/fiches/a_A090.htm, assessed on 15 July 2009.

OIE, Office Internationale d'Epizooties. 2006. Animal production food safety challenges in global markets. Revue Scientifique & Technique de l'Office Internationale des Epizooties, Paris, France. Vol. 25 (2).

Peters, A.R., Ball, J.P. 1995. Reproduction in Cattle. 2nd Edition. Blackwell Science, Osney Mead, Oxford, UK.

Philippe, P., Raboisson, D., Schelcher, F. 2008. Effects of heat treatment of colostrum. Le Nouveau Praticien Vétérinaire 167: 79-82.

Pierson, M. 1995. An overview of HACCP and its applications to animal production food safety. Symposium on Hazard Analysis Critical Control Points. Conference of Research Workers in Animal Diseases, November 12th, 1995. Chicago, IL, USA.

Rabin, M. 1998. Psychology and Economics. Journal of Economic Literature 36(1): 11-46.

Radostits, O.M., Leslie, K., Fetrow, J. 1994. Herd Health, Food Animal Production Medicine. 2nd Edition. W.B. Saunders Company Ltd, Philadelphia, PA, USA.

Radostits, O.M., Gay, C.C., Blood, D.C., Hinchcliff, K.W. 2000. Veterinary Medicine: A textbook of the disease of cattle, sheep, pigs, goats and horses. 9th Edition. W.B. Saunders Company Ltd, Philadelphia, PA, USA.

Radostits, O.M. 2001a. Diarreia aguda indiferenciada dos animais recém nascidos: Criptosporidiose dos vitelos. (*Undifferentiated acute diarrhoea of newborn animals, Cryptosporidiosis*). Revista Portuguesa de Buiatria, 1(December): 4-16.

Radostits, O.M. 2001b. Exame neurológico dos bovinos e diagnóstico diferencial de algumas doencas comuns do sistema nervoso dos bovinos: exame clinico em bovinos. (*Neurological examination of cattle and differential diagnosis of some common diseases of the bovine nervous system).* Revista Portuguesa de Buiatria, 1(December): 28-54.

Radostits, O.M., Gay, C.S., Hinchcliff, K.W., Constable, P.D. 2007. Veterinary Medicine. 10th Edition. Saunders/Elsevier, Philadelphia, PA, USA.

Richardson, E.K.B., Mee, J.F., Sanchez-Miguel, C., Crilly, J., More, S.J. 2009. Demographics of cattle positive for *Mycobacterium avium* subspecies *paratuberculosis* by faecal culture, from submissions to the Cork Regional Veterinary Laboratory. Irish Veterinary Journal 62: 398-405.

Rosenbaum-Nielsen, L. 2007. Biosecurity in cattle herds. In: Proceedings of the Buiatrie Francaise, European Meeting, Paris, France, 15-16 November 2007, pp. 43-45.

RVL. 2008. Regional Veterinary Laboratories – Surveillance Report 2008, Irish Department of Agriculture, Fisheries and Food, pp.1-28.

Ryan, D. 1997. Three HACCP-based programmes for quality management in cattle in Australia. Dairy Extension, NSW Australia.

Rydell, J. (editor). 2002/2008. A guide to calf milk replacers (types, use and quality). Publication of the Bovine Alliance on Management and Nutrition, revised 2002 and 2008, AFIA, Arlington VI, USA.

Schuijt, G., Taverne, M.A. 1994. The interval between birth and sternal recumbency as an objective measure of the vitality of newborn calves. Veterinary Record 135: 111-115.

Sellers, R. (editor). 2001. A guide to colostrum and colostrum management for dairy calves. Publication of the Bovine Alliance on Management and Nutrition. AFIA, Arlington, VI USA.

Sibley, R. 2006. Developing health plans for the dairy herd. In Practice 28: 114-121.

Smith, B.P. 2009. Large Animal Internal Medicine. 4th Edition. Mosby/Elsevier, St.Louis, Missouri, USA, pp. 248-251.

Staufenbiel, R., Schröder, U., Gelfert, C.C., Panicke, L. 2003. Körperkondition und Stoffwechselstabilität als Grundlage für eine hohe Milchleistung bei ungestörter Fruchtbarkeit und allgemeiner Gesundheit von Milchkühen. Archives Tierzucht. 46, 513-526.

Svensson, C., Linder, A., Olsson, S.-O. 2006. Mortality in Swedish dairy calves and replacement heifers. Journal of Dairy Science 89: 4769-4777.

Tesh, V.L., O'Brien, A.D., 1999. The pathogenic mechanisms of Shiga toxin and the Shiga-like toxins. Molecular Microbiology 5: 1817-1822.

Thrusfield, M., Ortega, C., De Blas, I., Noordhuizen, J.P.T.M., Frankena, K. 2001. WinEpiscope, improved epidemiological software for veterinary medicine. Veterinary Record 148: 567-572.

Towery, D. (editor). 2000. An introduction to infectious disease control and biosecurity on dairy farms. Publication of the Bovine Alliance on Management and Nutrition, published by AFIA, Arlington VI, USA.

Tozer, P.R., Heinrichs, A.J. 2001. What affects the costs of raising replacement heifers: a multiple component analysis. Journal of Dairy Science 84(8): 1836-1844.

Vaarst, M., Sorensen, J.T. 2009. Danish dairy farmers' perceptions and attitudes related to calf-management in situations of high versus no calf mortality. Preventive Veterinary Medicine 89: 128-133.

VanderFels-Klerx, H.J., Horst, S.S., Dijkhuizen, A.A. 2000. Risk factors for bovine respiratory disease in dairy youngstock in the Netherlands: the perception of experts. Livestock Production Science 66: 35-46.

Van Trierum, G. 2005. How efficient is calf rearing on your farm? International Dairy Topics 4(2): 16-17.

Vonk Noordegraaf, A., Buijtels, J.A., Dijkhuizen, A.A., Franken, P., Stegeman, J.A., Verhoeff, J. 1998. An epidemiological and economic simulation model to evaluate the spread and control of IBR in the Netherlands. Preventive Veterinary Medicine 36(3): 219-238.

Waltner-Toews, D., Martin, S.W., Meek, A.H. 1986. Dairy calf management, morbidity and mortality in Ontario Holstein herds III – association of management with morbidity. Preventive Veterinary Medicine 4: 137-156.

Weber, M.F., Groenendaal, H., Van Roermund, H.J., Nielen, M. 2004. Simulation of alternatives for the Dutch Johne's disease certification-and-monitoring program. Preventive Veterinary Medicine 62(1): 1-17.

Webster, J. 1995.Animal Welfare: a cool eye towards Eden. Blackwell Science Publishers, Oxford, UK, 273 pp.

Wolf, G. 1997. BVD/MD als Herdenproblem. ITB-Schriftenreihe, Verlag Hieronymus, Munchen, Germany.

Youngquist, R.S., Threlfall, W.R. 2007. Large Animal Theriogenology. W.B. Saunders Company, St.Louis, MO, USA, pp. 258, 270-473, 489.

Zaaijer, D., Noordhuizen, J.P.T.M. 2003. A novel scoring system for monitoring the relationship between nutritional efficiency and fertility in dairy cows. Irish Veterinary Journal 56: 145-151.

Websites which we advise for further consulting:

http://healthmap.org/r/OOTT

http://www.efsa.europa.eu/cs/BlobServer/Scientific_Opinion/biohaz_op_1189_dairy_cows_en.pdf?ssbinary=true

htpp://www.efsa.europa.eu/etc/medialib/efsa/science/colloquium_series/no4_animal_diseases/1179.Par.0017/File.dat/ses_summary_report_coll4_en1.pdf

http://www.focus.de/magazin/archiv/jahrgang_2009/ausgabe_40/

www.agri-ed.com/catalog.html

www.agri-ed.com/heifer.html

www.aphis.usda.gov/animal_health/

www.cdc.gov/ncidod/dbmd/

www.DQACenter.org/university

www.fao.org

www.nahms.usda.gov/

www.oie.org

www.partnersinreproduction.com

www.vacqa-international.com

www.vetvice.com

- subscribe to vph-l@mailserv.fao.org through mail to Francesco.Proscia@fao.org

Examples of software for HACCP applications:

- doHACCP by Norback, Ley & Associates LLC, 3022 Woodland Trail, Middleton, Wisconsin USA. www.norbackley.com
- QSA software Ltd. PO Box 306, St.Albans, Herts AL 1 3 DW, United Kingdom, HACCP (demo)software packages through www.qsa.co.uk

A simple and easy, public-domain software program for applied veterinary epidemiology:

- WINEPISCOPE (see also literature reference of Thrusfield *et al.*, 2001) available for free at either www.zod.wau.nl/qve or at http://winepi.unizar.es or at http://solismail.uu.nl
- Calculate *yourself* the sensitivity and specificity of a diagnostic test; estimate the sample size needed for answering a specific question; calculate the odds ratios for testing risk associations between a certain factor and a disease.

Keyword index

Printed in the United States
by Baker & Taylor Publisher Services